Arithmetic and Logic in Computer Systems

Mi Lu
Texas A&M University

A JOHN WILEY & SONS, INC., PUBLICATION

Copyright © 2004 by John Wiley & Sons, Inc. All rights reserved.

Published by John Wiley & Sons, Inc., Hoboken, New Jersey.
Published simultaneously in Canada.

No part of this publication may be reproduced, stored in a retrieval system or transmitted in any form or by any means, electronic, mechanical, photocopying, recording, scanning or otherwise, except as permitted under Section 107 or 108 of the 1976 United States Copyright Act, without either the prior written permission of the Publisher, or authorization through payment of the appropriate per-copy fee to the Copyright Clearance Center, Inc., 222 Rosewood Drive, Danvers, MA 01923, (978) 750-8400, fax (978) 646-8600, or on the web at www.copyright.com. Requests to the Publisher for permission should be addressed to the Permissions Department, John Wiley & Sons, Inc., 111 River Street, Hoboken, NJ 07030, (201) 748-6011, fax (201) 748-6008.

Limit of Liability/Disclaimer of Warranty: While the publisher and author have used their best efforts in preparing this book, they make no representation or warranties with respect to the accuracy or completeness of the contents of this book and specifically disclaim any implied warranties of merchantability or fitness for a particular purpose. No warranty may be created or extended by sales representatives or written sales materials. The advice and strategies contained herein may not be suitable for your situation. You should consult with a professional where appropriate. Neither the publisher nor author shall be liable for any loss of profit or any other commercial damages, including but not limited to special, incidental, consequential, or other damages.

For general information on our other products and services please contact our Customer Care Department within the U.S. at 877-762-2974, outside the U.S. at 317-572-3993 or fax 317-572-4002.

Wiley also publishes its books in a variety of electronic formats. Some content that appears in print, however, may not be available in electronic format.

Library of Congress Cataloging-in-Publication Data is available.

ISBN 0-471-46945-9

Printed in the United States of America.

10 9 8 7 6 5 4 3 2 1

To the memory of my mother, Shu Sheng Fan.
To my father, Chong Pu Lu, my husband, Jiming Yin, and my son, Luke Yin.

Contents

Preface		xiii
List of Figures		xv
List of Tables		xix
About the Author		xxi
1	**Computer Number Systems**	**1**
	1.1 Conventional Radix Number System	2
	1.2 Conversion of Radix Numbers	4
	1.3 Representation of Signed Numbers	7
	1.3.1 Sign-Magnitude	8
	1.3.2 Diminished Radix Complement	8
	1.3.3 Radix Complement	8
	1.4 Signed-Digit Number System	11
	1.5 Floating-Point Number Representation	15
	1.5.1 Normalization	15
	1.5.2 Bias	16
	1.6 Residue Number System	22
	1.7 Logarithmic Number System	23
	References	24
	Problems	26

viii CONTENTS

2	**Addition and Subtraction**	**29**
2.1	Single-Bit Adders	29
	2.1.1 Logical Devices	29
	2.1.2 Single-Bit Half-Adder and Full-Adders	32
2.2	Negation	35
	2.2.1 Negation in One's Complement System	36
	2.2.2 Negation in Two's Complement System	38
2.3	Subtraction through Addition	40
2.4	Overflow	43
2.5	Ripple Carry Adders	44
	2.5.1 Two's Complement Addition	44
	2.5.2 One's Complement Addition	46
	2.5.3 Sign-Magnitude Addition	48
	References	50
	Problems	52
3	**High-Speed Adder**	**53**
3.1	Conditional-Sum Addition	53
3.2	Carry-Completion Sensing Addition	56
3.3	Carry-Lookahead Addition (CLA)	61
	3.3.1 Carry-Lookahead Adder	61
	3.3.2 Block Carry Lookahead Adder	62
3.4	Carry-Save Adders (CSA)	66
3.5	Bit-Partitioned Multiple Addition	71
	References	73
	Problems	74
4	**Sequential Multiplication**	**77**
4.1	Add-and-Shift Approach	78
4.2	Indirect Multiplication Schemes	81
	4.2.1 Unsigned Number Multiplication	81
	4.2.2 Sign-Magnitude Number Multiplication	81
	4.2.3 One's Complement Number Multiplication	81
	4.2.4 Two's Complement Number Multiplication	85
4.3	Robertson's Signed Number Multiplication	87
4.4	Recoding Technique	89
	4.4.1 Non-overlapped Multiple Bit Scanning	89
	4.4.2 Overlapped Multiple Bit Scanning	90

	4.4.3	Booth's Algorithm	93
	4.4.4	Canonical Multiplier Recoding	95
	References		99
	Problems		100

5 Parallel Multiplication — 103
 5.1 Wallace Trees — 103
 5.2 Unsigned Array Multiplier — 105
 5.3 Two's Complement Array Multiplier — 108
  5.3.1 Baugh-Wooley Two's Complement Multiplier — 111
  5.3.2 Pezaris Two's Complement Multipliers — 117
 5.4 Modular Structure of Large Multiplier — 120
  5.4.1 Modular Structure — 120
  5.4.2 Additive Multiply Modules — 123
  5.4.3 Programmable Multiply Modules — 125
 References — 130
 Problems — 132

6 Sequential Division — 135
 6.1 Subtract-and-Shift Approach — 135
 6.2 Binary Restoring Division — 138
 6.3 Binary Non-Restoring Division — 141
 6.4 High-Radix Division — 144
  6.4.1 High-Radix Non-Restoring Division — 144
  6.4.2 SRT Division — 146
  6.4.3 Modified SRT Division — 147
  6.4.4 Robertson's High-Radix Division — 147
 6.5 Convergence Division — 150
  6.5.1 Convergence Division Methodologies — 152
  6.5.2 Divider Implementing Convergence Division Algorithm — 155
 6.6 Division by Divisor Reciprocation — 157
 References — 162
 Problems — 164

7 Fast Array Dividers — 167
 7.1 Restoring Cellular Array Divider — 167
 7.2 Non-Restoring Cellular Array Divider — 171

7.3	Carry-Lookahead Cellular Array Divider	173
	References	180
	Problems	181

8 Floating Point Operations — 183
8.1 Floating Point Addition/Subtraction — 183
8.2 Floating Point Multiplication — 184
8.3 Floating Point Division — 188
8.4 Rounding — 189
8.5 Extra Bits — 191
References — 194
Problems — 196

9 Residue Number Operations — 199
9.1 RNS Addition, Subtraction and Multiplication — 199
9.2 Number Comparison and Overflow Detection — 200
    9.2.1 Unsigned Number Comparison — 200
    9.2.2 Overflow Detection — 202
    9.2.3 Signed Numbers and Their Properties — 202
    9.2.4 Multiplicative Inverse and the Parity Table — 203
9.3 Division Algorithm — 206
    9.3.1 Unsigned Number Division — 206
    9.3.2 Signed Number Division — 209
    9.3.3 Multiplicative Division Algorithm — 212
References — 216
Problems — 218

10 Operations through Logarithms — 221
10.1 Multiplication and Addition in Logarithmic Systems — 221
10.2 Addition and Subtraction in Logarithmic Systems — 222
10.3 Realizing the Approximation — 225
References — 232
Problems — 233

11 Signed-Digit Number Operations — 235
11.1 Characteristics of SD Numbers — 235
11.2 Totally Parallel Addition/Subtraction — 236
11.3 Required and Allowed Values — 237

11.4	*Multiplication and Division*	*239*
	References	*243*
	Problems	*244*

Index   **245**

Preface

This book describes the fundamental principles of computer arithmetic. Algorithms for performing operations like addition, subtraction, multiplication and division in digital computer systems are presented. The goal is to explain the concepts behind the algorithms rather than to address any direct applications. Alternative methods are examined and various possibilities considered. With the rapid growth of VLSI technology, some currently unattractive algorithms may be implemented with remarkable performance in the future.

This book can be used as a text of an introductory course for graduate students or senior undergraduate students in electrical engineering, and computer and mathematical sciences. It can also be used as a reference book for practicing engineers and computer scientists involved in the design, application and development of computer arithmetic units. For the number systems covered in Sections 1.4, 1.6 and 1.7, some exercise problems are listed in Chapters 9, 10 and 11 for in-depth study.

I have been teaching a computer arithmetic course for fifteen years and have supervised Doctorate and Masters research projects in this area. As a preliminary version of the book, my lecture notes have received positive and constructive feedback over the years. An effort has been made to keep fundamental material self-contained and instructive rather than just referring readers to articles spread throughout the literature. The theories in the book have been carefully derived and the reasoning addressed as completely as possible. In addition to "it is so," pointed to the readers is "why it is so." The notation in different discussions is unified and the descriptions are given logically and with clarity. The whole presentation of the text is designed to be smooth and coherent rather than a collection of broken pieces, with leaps from one subject to another. I gratefully thank my father, Chong Pu Lu, and my husband, Jiming Yin, for their encouragement and support during the writing of this book. I also wish to acknowledge the contribution made by my graduate student, C. T. Chiang, for his assistance in graphical typesetting.

College Station, Texas MI LU

List of Figures

1.1	Floating-Point Representation	18
1.2	Range of the Numbers	19
1.3	Precision of Floating-Point Numbers	19
1.4	Double Precision Floating-Point Representation	20
2.1	AOI Function	30
2.2	Decoder and Multiplexer	31
2.3	Single-Bit Half-Adder	32
2.4	Design of Full-Adder	34
2.5	Single-Bit Subtrator	36
2.6	Negation in One's Complement System	37
2.7	Negation in Two's Complement System	39
2.8	Subtraction through Addition	41
2.9	One-Bit Adder/Subtractor	42
2.10	Two's Complement Addition/Subtraction	45
2.11	One's Complement Addition/Subtraction	47

2.12	Block Diagram of Sign-Magnitude Addition/Subtraction	49
2.13	Sign-Magnitude Addition/Subtraction	50
3.1	Conditional-Sum Addition	55
3.2	Conditional-Sum Adder	57
3.3	Generation and Transmission of Carries	58
3.4	Construction of Carry-Completion Sensing Adder	59
3.5	Carry-Lookahead Adder	63
3.6	Block Carry-Lookahead Adder	65
3.7	Carry-Save Adder	67
3.8	Carry-Save Adder Tree	69
3.9	Two Types of Parallelization in Multi-Operand Addition	70
3.10	Bit-Partitioned Multiple Addition	72
3.11	Carry-Completion Sensing Adder	75
3.12	Carry-Save Adder	75
3.13	Bit-Partitional Adder	76
4.1	Hardware for Sequential Multiplication	79
4.2	Register Occupation	80
4.3	Unsigned Number Multiplication	82
4.4	Sign-Magnitude Number Multiplication	83
4.5	One's Complement Number Multiplication	84
4.6	Two's Complement Number Multiplication	86
4.7	Negative Multiplicand Times Positive Multiplier	87
4.8	Negative Multiplicand Times Negative Multiplier	88
4.9	Multiple Bit Scanning	90
4.10	String Property	91
4.11	Two-Bit Scan vs. Overlapped Three-Bit Scan	92
4.12	Example of Booth's Multiplication	94
4.13	Scan Pattern in 32-bit Multiplication	97

4.14	Adding the Bit-Pairs Parallelly Scanned with a CSA Tree	98
5.1	Wallace Tree	104
5.2	5-by-5 Multiplication	106
5.3	5×4 Array Multiplier Performing 5-by-5 Multiplication	108
5.4	Different Types of Full Adders	111
5.5	Distribution of Negative Weight	114
5.6	Baugh-Wooley Array Multiplier Performing 6-by-4 Two's Complement Multiplication	115
5.7	Baugh-Wooley Multiplication for $10 \times (-3)$	115
5.8	Baugh-Wooley Array with $m=n=5$	116
5.9	Distribution of the Negative Weight	117
5.10	5-by-5 Pezaris Array Multiplier	118
5.11	The Adjustment	118
5.12	5-by-5 Bi-Section Array Multiplier	119
5.13	5-by-5 Tri-section Array Multiplier	120
5.14	Alignment of the Sub-Products	121
5.15	8-by-8 Multiplication via 4-by-4 Multipliers	121
5.16	Modular Structure of Array Multipliers	122
5.17	4-by-2 Additive Multiply Module	123
5.18	8-by-8 Multiplication via 4-by-2 Multipliers	124
5.19	Modular Structure Applying Additive Multiply Modules	125
5.20	Combine Small AMMs into a Large One	127
5.21	Summands of Preparation in Programmable AMM	128
5.22	AMM 8×8 Applying AMM 4×4	129
6.1	Pencil-and-Paper Division	136
6.2	Long Division Form	137
6.3	Example of Long Division	138
6.4	Example of Restoring Procedure	139

6.5	Hardware for Restoring Division	140
6.6	Division Performed by Non-Restoring/Restoring Algorithms	143
6.7	Flow Chart for Wilson-Ledley's Division Algorithm	148
6.8	Numerical Example for Wilson-Ledley's Division Algorithm	149
6.9	Robertson Diagrams	151
6.10	Stepwise Approximation of the Reciprocal of Divisor	160
7.1	4-by-4 Restoring Array Divider	168
7.2	5-by-5 Non-Restoring Array Divider	172
7.3	Carry-Lookahead Array Divider for 4-bit Division (Carry-Lookahead Mechanism is Shown in the Second Row Only)	175
7.4	Example of Carry-Lookahead Array Division	178
7.5	Wires Can Take Up Significant Space	179
8.1	Data Flow of Floating Point Addition/Subtraction	185
8.2	Data Flow of Floating Point Multiplication	187
8.3	Data Flow of Floating Point Division	189
8.4	Example of Rounding in Subtraction	194
9.1	Flowchart of the Unsigned Number Division Algorithm	210
9.2	Example of Signed Number Division	211
9.3	Example of Conversion to Mixed-Radix Representation	214
10.1	Linear Approximation of $\log_2(1+x)$	224
10.2	Mechanism for Multiplication (Division) in Binary Logarithms	225
10.3	Logarithmic Curve and Four-Straight-Line Approximation	227
10.4	Error of the Four-Straight-Line Approximation	228
10.5	Correction Register	229
10.6	Realization of the Correction	230
11.1	Totally-Parallel Adder in Signed-Digit System	237

List of Tables

1.1	Numbers Represented by 4 bits in Different Number Systems	12
1.2	Finding Signed Digits	14
1.3	Reserved Representation in IEEE Standard	21
2.1	Delay Time and Area of Logic Gates	30
2.2	Logic Function of a Half-Adder	32
2.3	Logic Function of a Full-Adder	33
2.4	Single-Bit Subtractor	35
2.5	Negation in One's Complement System	36
3.1	Maximum Inputs of CSA Trees	71
4.1	Recoding the Triplets	92
5.1	Combination and Delay of k-input Wallace Tree	105
6.1	2-Input 4-Output ROM to Store $p_0(s)$.	158
8.1	Round to Nearest Even	191
9.1	Parity Table for Modulus Set $\{3, 5, 7\}$	201

9.2	Mixed-Radix Digits	213
10.1	Required $\frac{1}{2^F}$s.	229
10.2	Mean-Square Error and Coefficients for Logarithm Approximation	231
10.3	Logarithm Equations	232
11.1	Example for SD Multiplication	241
11.2	Example for SD Division	242

About the Author

Mi Lu received the M.S. and Ph.D. degrees in electrical engineering from Rice University, Houston, in 1984 and 1987, respectively. She joined the Department of Electrical Engineering at Texas A&M University in 1987, where she is currently a professor. Lu's research interests include computer arithmetic, parallel computing, computer architectures, VLSI algorithms and computer networks, and she has published more than 100 technical papers in these areas. In addition, Professor Lu has served as associate editor of the *Journal of Computing and Information* and the *Information Sciences Journal*, and was conference chairman of the Fifth, Sixth and Seventh International Conferences on Computer Science and Informatics. She served on the panel of the National Science Foundation and the panel of the IEEE Workshop on Imprecise and Approximate Computation, as well as many conference program committees. Professor Lu is also the chairman of 60 research advisory committees for Ph.D. and Masters students, is a registered professional engineer, and is a senior member of the Institute of Electrical and Electronics Engineers. She is recognized in *Who's Who in the World* (2001, 2003), *Who's Who in America* (2002–2003) and *Who's Who of American Women* (2002–2003).

1
Computer Number Systems

As the arithmetic applications grow rapidly, it is important for computer engineers to be well informed of the essentials of computer number systems and arithmetic processes.

With the remarkable progress in the very large scale integration (VLSI) circuit technology, many complex circuits unthinkable yesterday become components easily realizable today. Algorithms that seemed impossible to implement now have attractive implementation possibilities for the future. This means that not only the conventional computer arithmetic methods, but also the unconventional ones are worth investigation in new designs.

Numbers play an important role in computer systems. Numbers are the basis and object of computer operations. The main task of computers is computing, which deals with numbers all the time.

Humans have been familiar with numbers for thousands of years, whereas representing numbers in computer systems is a new issue. A computer can provide only finite digits for a number representation (fixed word length), though a real number may be composed of infinite digits.

Because of the tradeoffs between word length and hardware size, and between propagation delay and accuracy, various types of number representation have been proposed and adopted. In this book, we introduce the Conventional Radix Number System and Signed-Digit Number System, both belonging to the Fixed-Point Num-

1.1 CONVENTIONAL RADIX NUMBER SYSTEM

A *conventional radix number* N can be represented by a string of n digits such as

$$(d_{n-1}d_{n-2}\cdots d_1 d_0)_r$$

with r being the radix. d_i, $0 \leq i \leq n-1$, is a digit and $d_i \in \{0, 1, \cdots, r-1\}$. Note that the position of d_i matters, such as 27 is a different number from 72. Such a number system is referred to as a *positional weighted* system. Actually,

$$\begin{aligned} N &= d_{n-1} \cdot w_{n-1} + d_{n-2} \cdot w_{n-2} + \cdots + d_0 \cdot w_0 \\ &= \sum_{i=0}^{n-1} d_i \cdot w_i \end{aligned}$$

with w_i being the weight of position i. If r is fixed, as in the *fixed-radix number system* in our further discussion, $w_i = r^i$. Hence,

$$N = d_{n-1} \cdot r^{n-1} + d_{n-2} \cdot r^{n-2} + \cdots + d_0 \cdot r^0 \quad (1.1)$$

$$= \sum_{i=0}^{n-1} d_i \cdot r_i. \quad (1.2)$$

If r is not fixed, the number becomes a *mixed-radix number*. For example, to represent time T we have $T = [hour : minute : second]$ or $T = [(h)_{r2}, (m)_{r1}, (s)_{r0}]$ where $r_2 = 24$; $r_1 = 60$; $r_0 = 60$.

To include the fraction into a fixed radix number N, let "." be a radix point with the integer part on the left of it and fraction part on the right of it. There are n digits in the integer and k digits in the fraction, such as

$$(d_{n-1}\cdots d_0.d_{-1}\cdots d_{-k})_r.$$

Then

$$N = \sum_{i=-k}^{n-1} d_i \cdot r^i. \quad (1.3)$$

For example, in the decimal number system, $r = 10$, and $d_i \in \{0, 1, \cdots, 9\}$.

$$\begin{aligned} N &= (69.3)_{10} \\ &= d_1 \cdot r^1 + d_0 \cdot r^0 + d_{-1} \cdot r^{-1} \\ &= 6 \times 10^1 + 9 \times 10^0 + 3 \times 10^{-1} \\ &= 69.3. \end{aligned}$$

In the octal number system, $r = 8$, and $d_i \in \{0, 1, \cdots, 7\}$.

$$\begin{aligned} N &= 47.2_8 \\ &= d_1 \cdot r^1 + d_0 \cdot r^0 + d_{-1} \cdot r^{-1} \\ &= 4 \times 8^1 + 7 \times 8^0 + 2 \times 8^{-1} \\ &= 32 + 7 + 0.25 = 39.25. \end{aligned}$$

In the hexadecimal number system, $r=16$. Capital letters A through F are used to represent the numbers 10 through 15.

$$\begin{aligned} N &= 2A.C_{16} \\ &= d_1 \cdot r^1 + d_0 \cdot r^0 + d_{-1} \cdot r^{-1} \\ &= 2 \times 16^1 + 10 \times 16^0 + 12 \times 16^{-1} \\ &= 32 + 10 + 0.75 = 42.75. \end{aligned}$$

In the binary number system, $r=2$, and $d_i \in \{0, 1\}$.

$$\begin{aligned} N &= (10.1)_2 \\ &= d_1 \cdot r^1 + d_0 \cdot r^0 + d_{-1} \cdot r^{-1} \\ &= 1 \times 2^1 + 0 \times 2^0 + 1 \times 2^{-1} \\ &= 2 + 0.5 = 2.5. \end{aligned}$$

In the string of weighted digits $(d_{n-1} \cdots d_0.d_{-1} \cdots d_{-k})_r$, d_{n-1} is called the most significant digit (MSD), and d_{-k} the least significant digit (LSD). A binary digit is referred to as a bit, and the above two digits are MSB and LSB, respectively. In an electric circuit, there are two voltage levels, "high" and "low", which can easily represent two digits, "1" and "0", in the binary number system. Of course, more bits are required to represent a number in binary than in other radix systems. Remember, the number of bits required to encode a number λ is $\lfloor log_2 \lambda \rfloor + 1$.

Here the *downstile* or *floor* of x $\lfloor x \rfloor$, is the greatest integer that is not greater than x, where x can be an integer or real. (Likewise, the *upstile* or *ceiling* of x $\lceil x \rceil$, is the smallest integer that is not smaller than x.)

For example, to represent the decimal number 10, the number of bits required is

$$\lfloor log_2 10 \rfloor + 1 = \lfloor 3.322 \rfloor + 1 = 4.$$

4 COMPUTER NUMBER SYSTEMS

1.2 CONVERSION OF RADIX NUMBERS

While computer systems recognize the binary, octal and hexadecimal numbers, humans are most familiar with decimal number systems. Numbers can be converted from one radix system to another before, after or in the middle of arithmetic operations. We present below the algorithms for such conversions.

Given an integer,

$$(d_{n-1}d_{n-2}\cdots d_1 d_0)_r,$$

with base r other than 10, such as $r = 2$ in binary, $r = 8$ in octal or $r = 16$ in hexadecimal, according to Equation (1.2), the following equation provides a method to convert it to the corresponding decimal number N_1.

$$N_1 = d_{n-1} \cdot r^{n-1} + d_{n-2} \cdot r^{n-2} + \cdots + d_0 \cdot r^0. \tag{1.4}$$

That is, N_1 can be obtained by performing the multiplication of each given digit, the weight it carries and summing all the products.

In the reversed way, given a decimal number we can obtain the corresponding digits in its binary, octal or hexadecimal representation by division, using r as the divisor equal to 2, 8 or 16, respectively.

Dividing both sides of Equation (1.4) by r, we have on the right-hand side the remainder d_0 and the quotient

$$d_{n-1} \cdot r^{n-2} + d_{n-2} \cdot r^{n-3} + \cdots + d_1,$$

since $d_0 < r$ and other terms on the right-hand side are integer times of r. If we divide the above quotient again by r, we will obtain the remainder d_1, and so forth. After performing the division $n - 1$ times, d_{n-1} will become the quotient. If we divide it by r again, we will have quotient 0, since any $d_i < r$ and the last remainder d_{n-1}. The conversion procedure stops there.

So, to convert a decimal integer to a radix r number one can let the decimal number be the initial quotient, repeatedly divide the quotient by r and record the remainder until the quotient is zero. Then write the digits in the radix r number from left to right using the sequence of remainders, last obtained first. Following is a numerical example to show how the decimal number 10 is converted to the binary number $(1010)_2$. Here $r = 2$ is the number we should repeatedly divide by. We write the quotient below the given number 10 and the remainder on the right of the quotient.

$$\begin{array}{r} \underline{|\ 10} \\ \underline{|\ 5} \cdots 0 \end{array}$$

$$\begin{array}{r|l}\underline{}&2\cdots 1\\\underline{1}&1\cdots 0\\0&\cdots 1\end{array}.$$

For a radix r fraction number

$$(0.d_{-1}d_{-2}\cdots d_{-k})_r,$$

with $r \neq 10$, the corresponding decimal number N_2 can be obtained by

$$N_2 = d_{-1} \cdot r^{-1} + d_{-2} \cdot r^{-2} + \cdots + d_{-k} \cdot r^{-k}. \tag{1.5}$$

On the other hand, a decimal fraction can be converted to a radix r number such as a binary, octal or hexadecimal number with r being 2, 8 or 16, respectively.

Multiplying both sides of Equation (1.5) by r, we have on the right-hand side

$$d_{-1} + d_{-2} \cdot r^{-1} + \cdots + d_{-k} \cdot r^{-k+1},$$

where d_{-1} is the integer part and others add up to the fraction part. Multiply the fraction part by r again, we have

$$d_{-2} + d_{-3} \cdot r^{-1} + \cdots + d_{-k} \cdot r^{-k+2},$$

where d_{-2} is the integer part. Repeating the multiplication process and retaining the digits in the integer part, we can obtain the radix r number corresponding to the given decimal fraction. If we stop when the fraction part becomes zero, then the conversion completes precisely. Note that the fraction part may never become zero. Then, depending on how many digits are allowed in the radix r number, one can decide the time to stop the multiplication procedure. Infinite digits may be required to represent the given decimal fraction precisely in the radix r number system. With a limited number of digits, the found radix r number is an approximation.

So, to convert a decimal fraction to a radix r number, we can repeatedly multiply the fraction part by r and retain the integer digit in sequence until the fraction is zero, or the number of digits required are obtained. Then we write the digits following the radix point from left to right with the integer digits, the earliest obtained first.

Following is an example to show how the decimal number 0.5625 is converted to the binary number $(0.1001)_2$. Here, $r = 2$ is the number we should repeatedly multiply by. We put the integer digit in parentheses after each multiplication and it won't participate in the next multiplication except the fraction part.

$$\begin{array}{r}0.5625\\\times2\\\hline(1).1250\end{array}$$

$$\begin{array}{r} \times \quad 2 \\ \hline (0).2500 \\ \times \quad 2 \\ \hline (0).5000 \\ \times \quad 2 \\ \hline (1).0000 \,. \end{array}$$

As to the conversion between binary numbers and octal numbers, or between binary and hexadecimal numbers, the following has been observed.

In octal numbers $r = 8$ and digit $0 \leq d_i \leq 7$. In binary numbers $r = 2$ and $8 = 2^3$. 3 bits in binary are necessary and sufficient to represent the value of one digit in octal. For example, $(5)_8 = (101)_2$ and $(7)_8 = (111)_2$. Hence, to convert an octal number to a binary number, represent each digit in octal with 3 bits in binary, and concatenate all the bits together. For example,

$$= \underbrace{100}_{4} \underbrace{111}_{7} \underbrace{010}_{2} \underbrace{110}_{6} \,.\, \underbrace{001}_{1_8} {}_2 \,.$$

On the other hand, to convert a binary number to an octal, group each 3 bits together starting from the radix point. For the integer part, group from right to left, and add 0(s) on the left if the last group contains less than 3 bits. For the fraction part, group from left to right, and add 0(s) on the right if the last group contains less than 3 bits. Find the value for each group applying Equation (1.3).

In hexadecimal numbers $r = 16$ and digit $0 \leq d_i \leq 15$. In binary numbers $r = 2$ and $16 = 2^4$. 4 bits in binary are necessary and sufficient to represent the value of one digit in hexadecimal. For example, $(9)_{16} = (1001)_2$, and $(15)_{16} = (1111)_2$. Hence, to convert a hexadecimal number to a binary number, represent each digit in hexadecimal with 4 bits in binary and concatenate all the bits together.

On the other hand, to convert a binary number to a hexadecimal, group each 4 bits together starting from the radix point. For the integer part, group from right to left and add 0(s) on the left if the last group contains less than 4 bits. For the fraction part, group from left to right and add 0(s) on the right if the last group contains less than 4 bits. Find the value for each group applying Equation (1.3). For example,

$$\underbrace{10}_{= 2} \underbrace{1010}_{A} \underbrace{1001}_{9} \underbrace{1111}_{F} \,.\, \underbrace{1100}_{C_{16}} {}_2 \,.$$

In the decimal to octal (hexadecimal, respectively) conversion, instead of the dividing by 8 (by 16) operation described earlier, one can first convert the decimal number to binary since dividing by 2 is easier, and then convert the binary number to octal (hexadecimal).

On the other hand, in the octal (hexadecimal) to decimal conversion, instead of the multiplying by 8 (by 16) operation described earlier, one can first convert the octal

1.3 REPRESENTATION OF SIGNED NUMBERS

All the numbers referred to so far are unsigned numbers. As negative numbers are often involved in scientific computing, the representation of signed numbers is discussed below.

Let a conventional radix number A be an n digit signed number with the MSD representing the sign. That is,

$$A = (a_{n-1}a_{n-2}\cdots a_1 a_0)_r,$$

and the *sign digit* a_{n-1} is decided as follows:

$$a_{n-1} = \begin{cases} 0 & \text{if } A \geq 0 \\ r-1 & \text{if } A < 0. \end{cases}$$

Note that for an integer number, the radix point is on the right of a_0, that is,

$$(a_{n-1}a_{n-2}\cdots a_1 a_0.),$$

and for a fraction number, the radix point is on the left of a_{n-2}, such as

$$(a_{n-1}.a_{n-2}\cdots a_1 a_0).$$

Particularly, when $a_{n-2} \neq 0$, we say that A is a *normalized fraction*.

In the discussion below, we assume that A is an integer for illustration. Let the magnitude of A,

$$|A| = (m_{n-2}\cdots m_1 m_0).$$

If $a_{n-1} = 0$, A is a positive number. Then,

$$\begin{aligned} A &= (0a_{n-2}\cdots a_1 a_0)_r \\ &= (0m_{n-2}\cdots m_1 m_0)_r. \end{aligned}$$

That is, number A has the same value as its true magnitude.

$$A = \sum_{i=0}^{n-2} a_i r^i = \sum_{i=0}^{n-2} m_i r^i.$$

8 COMPUTER NUMBER SYSTEMS

If $a_{n-1} = r - 1$, A is a negative number, then the representation of the number will depend on which format is used.

There are three representations of a negative number: (1) sign-magnitude, (2) diminished radix complement, and (3) radix complement.

1.3.1 Sign-Magnitude

$$(r-1)m_{n-2}m_{n-3}\cdots m_1 m_0 .$$

Some examples of sign-magnitude representation are:

$$r = 2, \quad (1)1010 = -1010_2 = -10_{10},$$
$$r = 10, \quad (9)7602 = -7602_{10}.$$

1.3.2 Diminished Radix Complement

$$(r-1)\overline{m}_{n-2}\overline{m}_{n-3}\cdots \overline{m}_1 \overline{m}_0 ,$$

where

$$\overline{m}_i = (r-1) - m_i, \quad 0 \le i \le n-2.$$

The diminished radix complement representation is also known as $(r-1)$'s complement denoted as

$$\overline{A} = r^n - 1 - |A|,$$

where n is the total number of digits including the sign digit. For example, given $r = 2, \overline{A} = 2^n - 1 - |A|$, and we have the 1's complement representation as follows:

$$-1010_2 : \quad (1)0101.$$

Given $r = 10, \overline{A} = 10^n - 1 - |A|$, we have the following 9's complement representation.

$$-7602_{10} : \quad (9)2397.$$

1.3.3 Radix Complement

$$((r-1)\overline{m}_{n-2}\overline{m}_{n-3}\cdots \overline{m}_1 \overline{m}_0) + 1,$$

where

$$\overline{m_i} = (r-1) - m_i, \quad 0 \le i \le n-2.$$

The radix complement representation is also called r's complement, denoted as

$$\overline{A} = r^n - |A|.$$

For example, given $r = 2$, $\overline{A} = 2^n - |A|$, and we have the 2's complement representation as follows:

$$-1010_2 : \quad (1)0110.$$

Given $r = 10$, $\overline{A} = 10^n - |A|$, and we have the following 10's complement representation:

$$-7602_{10} : \quad (9)2398.$$

Next we discuss the representation of a fraction number. If

$$B = (0.p_{-1}p_{-2}\cdots p_{-k})_r,$$

B is a positive number. It has the same value as the true magnitude of B.

$$B = |B| = \sum_{j=-k}^{-1} p_j \cdot r^j.$$

Compare with Equation (1.3)

$$N = \sum_{i=-k}^{n-1} d_i \cdot r^i,$$

$n = 0$ here.

If

$$B = ((r-1).p_{-1}p_{-2}\cdots p_{-k})_r,$$

B is a negative number. Then B has the following representations:

(1) Sign-magnitude

$$(r-1).p_{-1}p_{-2}\cdots p_{-k}.$$

(2) Diminished radix complement

$$(r-1).\overline{p}_{-1}\overline{p}_{-2}\cdots\overline{p}_{-k},$$

where

$$\overline{p}_j = (r-1) - p_j, \quad -k \leq j \leq -1.$$

The diminished radix complement representation of B can be found by

$$\overline{B} = r^1 - r^{-k} - |B|.$$

(3) Radix complement

$$((r-1).\overline{p}_{-1}\overline{p}_{-2}\cdots\overline{p}_{-k}) + r^{-k},$$

where

$$\overline{p}_j = (r-1) - p_j, \quad -k \leq j \leq -1.$$

The radix complement representation of B can be found by

$$\overline{B} = r^1 - |B|.$$

In the previous discussion given a positive number, say $|A|$, the negative number $\overline{A} = -|A|$ is represented in the *complement form*. $\overline{A} + |A| = r^n - 1$ in the $(r-1)$'s complement system, or $\overline{A} + |A| = r^n$ in the r's complement system, thereby we say that $\overline{A}$ and $|A|$ are the *complement numbers of* each other. As we know, $\overline{A}$ is the complement number of $|A|$. Given a positive number $|A|$, by performing the *complement operation* $r^n - 1 - |A|$ or $r^n - |A|$, we can find the complement number of $|A|$, that is, $\overline{A} = -|A|$. On the other hand, $|A|$ is the complement number of $\overline{A}$. Given a negative number $\overline{A}$, we should be able to find the complement number of it, that is, $|A|$, by the similar complement operation. It can be seen that the complement operation is reciprocal.

To verify this let the given negative number $\overline{A}$ be in the $(r-1)$'s complement system. The representation of it should have already been in the complement form. That is,

$$(r-1)\overline{m}_{n-2}\overline{m}_{n-3}\cdots\overline{m}_1\overline{m}_0,$$

where

$$\overline{m}_i = (r-1) - m_i, \quad 0 \leq i \leq n-2.$$

Performing the $(r-1)$'s complement operation, we have

$$\begin{aligned} r^n - 1 - \overline{A} &= ((r-1)(r-1)(r-1)\cdots(r-1))_r - ((r-1)\overline{m}_{n-2}\overline{m}_{n-3}\cdots\overline{m}_0)_r \\ &= (0m_{n-2}m_{n-3}\cdots m_0)_r \\ &= |A|. \end{aligned}$$

Given n digits, the numbers that can be represented in different systems are as follows. In the sign-magnitude system, $00\cdots00$ represents 0. On the positive side, $00\cdots01$ to $0(r-1)\cdots(r-1)(r-1)$ represent positive numbers 1 to $r^{n-1}-1$ in an ascending order. On the negative side, $(r-1)0\cdots0$ represents -0. From $(r-1)0\cdots01$ to $(r-1)(r-1)\cdots(r-1)(r-1)$, negative numbers -1 to $-(r^{n-1}-1)$ are represented in a descending order.

In the $(r-1)$'s complement number system, 0 is represented by $00\cdots0$, and the $(r-1)$'s complement number of it is $(r-1)(r-1)\cdots(r-1)$ which means -0. The positive numbers represented are as same as in the sign-magnitude system. From $(r-1)0\cdots00$ to $(r-1)(r-1)\cdots(r-1)(r-2)$, negative numbers $-(r^{n-1}-1)$ to -1 are represented in an ascending order. In both the sign-magnitude system and the $(r-1)$'s complement system, the maximum number represented by n digits is $r^{n-1}-1$, and the smallest number represented is $-(r^{n-1}-1)$. In both systems there are two representations for value 0, $+0$ and -0.

In the r's complement number system there is only one representation for 0, that is, $00\cdots0$. The positive numbers represented are the same as in the other two systems. The maximum positive number represented is still $r^{n-1}-1$, denoted as $0(r-1)(r-1)\cdots(r-1)$. The r's complement number of it is $(r-1)0\cdots01$, which does not represent the smallest number however. $(r-1)0\cdots00$ is the smallest number which is 1 less than $(r-1)0\cdots01$. This number is not the complement of any positive number represented by the n digits, since the r's complement number of it is $(r-1)0\cdots00$ itself which is not a positive number. Hence, from $(r-1)0\cdots00$ to $(r-1)(r-1)\cdots(r-1)(r-1)$, negative numbers $-(r^{n-1})$ to -1 are represented in an ascending order. Compared with the other two number systems, one more negative number can be represented. The maximum number represented is $r^{n-1}-1$ and the smallest number represented is $-r^{n-1}$.

Given 4 digits in the binary system, Table 1.1 shows the numbers represented in the sign-magnitude, r's complement and $(r-1)$'s complement systems. It can be observed that the column for 1's complement seems symmetric, but that for the 2's complement is not.

1.4 SIGNED-DIGIT NUMBER SYSTEM

The number systems introduced in the previous sections belong to the conventional radix number system, which is non-redundant, positional and weighted. Each digit d_i has only a positive value, and is less than radix r. That is,

$$0 \leq d_i \leq r-1.$$

12 COMPUTER NUMBER SYSTEMS

Table 1.1: Numbers Represented by 4 bits in Different Number Systems

Binary bits	Sign-magnitude	1's complement	2's complement
0000	0	0	0
0001	1	1	1
0010	2	2	2
0011	3	3	3
0100	4	4	4
0101	5	5	5
0110	6	6	6
0111	7	7	7
1000	−0	−7	−8
1001	−1	−6	−7
1010	−2	−5	−6
1011	−3	−4	−5
1100	−4	−3	−4
1101	−5	−2	−3
1110	−6	−1	−2
1111	−7	−0	−1

If $d_i \geq r$ is allowed, then a given number $(\cdots a_{i+1} a_i \cdots)$ can always be rewritten as $(\cdots (a_{i+1} + 1)(a_i - r) \cdots)$, since

$$1 \cdot r^{i+1} = r \cdot r^i,$$

and the increased value caused by the $(i+1)$th digit update is equal to the decreased value caused by the ith digit update. In this case, a number can be represented in two or more forms and the representation is redundant. Hence, defining $0 \leq d_i \leq r - 1$ assured the non-redundant representation.

Next, we introduce a new number system, the signed-digit number system, in which each digit can have either a positive or a negative value and the representation is redundant.

Let a signed-digit number

$$X = (x_{n-1} \cdots x_0.x_{-1} \cdots x_{-k})_r.$$

Given radix $r \geq 2$, each digit of a Signed-Digit Number, x_i, has any value of $\{-\alpha, \cdots -1, 0, 1, \cdots \alpha\}$, that is,

$$-\alpha \leq x_i \leq \alpha,$$

where

$$\lceil \tfrac{r-1}{2} \rceil \leq \alpha \leq r - 1.$$

Since the signed-digit representation of a number may not be unique, we choose $\alpha = \lfloor \frac{r}{2} \rfloor$ for the minimum redundancy.

For example, if $r = 4$, $\alpha = \lfloor \frac{4}{2} \rfloor = 2$. Then digit $x_i \in \{-2, -1, 0, 1, 2\}$.

To find the value of X, we have

$$X = \sum_{i=-k}^{n-1} x_i \cdot r^i, \quad x_i \in \{-\alpha, \cdots -1, 0, 1, \cdots \alpha\}.$$

For example, given $r = 4$, $X_{SD} = (1\bar{2}.2)_4$, we have

$$\begin{aligned} X &= (1 \times 4^1) + (\bar{2} \times 4^0) + (2 \times 4^{-1}) \\ &= 4 + (-2) + (0.5) \\ &= 2.5. \end{aligned}$$

In general, X can be either positive or negative without putting a "sign" in front of it.

Given $r = 2$, $X_{SD} = (00\bar{1}\bar{1})_2$, $\alpha = \lfloor \frac{2}{2} \rfloor = 1$, $x_i \in \{-1, 0, 1\}$.

$$X = \bar{1} \times 2^1 + \bar{1} \times 2^0 = -3.$$

Note that given $X_{SD} = (0\bar{1}01)_2$

$$X = \bar{1} \times 2^2 + 1 \times 2 = -3.$$

Obviously, the signed-digit representation of a number of particular value may not be unique. For $X = -3$, $n = 4$, $k = 0$ and $r = 2$, we can have the signed-digit representation of -3 as

$$X = (00\bar{1}\bar{1})_2$$

or

$$X = (0\bar{1}01)_2$$

or

$$X = (\bar{1}101)_2$$

Table 1.2: Finding Signed Digits

i	4	3	2	1	0
x_i		0	6	4	8
b_i	0	1	0	1	
d_i		0	$\bar{4}$	4	$\bar{2}$
$y_i = b_i + d_i$		1	$\bar{4}$	5	$\bar{2}$

or
$$X = (0\bar{1}1\bar{1})_2$$
or
$$X = (\bar{1}11\bar{1})_2.$$

Among all the valid representations of a number, the *minimal signed-digit representation* is the one that has a minimal number of non-zero digits.

For $X = -3$, $(00\bar{1}\bar{1})_2$ and $(0\bar{1}01)_2$ are the minimal signed-digit representations.

We discuss below the conversion of a conventional radix number $X = (x_{n-1}, \cdots, x_1, x_0)_r$ to the signed-digit number $Y = (y_{n-1}, \cdots, y_1, y_0)_r$. Digit y_i can be obtained by

$$y_i = d_i + b_i,$$

where the *borrow digit* b_{i+1} can be decided as follows,

$$b_{i+1} = \begin{cases} 0 & \text{if } x_i < \alpha \\ 1 & \text{if } x_i \geq \alpha \end{cases},$$

and the *interim difference digit* d_i can be decided by

$$d_i = x_i - r \cdot b_{i+1}.$$

Let $X = (0648)_{10}$. Given $r = 10$, $n = 4$ and $\alpha = 6$, we have the digit set $\{\bar{6}, \cdots, \bar{1}, 0, 1, \cdots, 6\}$. The corresponding b_is, d_is and resulted y_is are listed in Table 1.2.

So, $Y = (1\bar{4}5\bar{2})_{10}$. To check the result, we find the total weight carried by the positive digits and subtract from it the total weight carried by the negative digits. That is, $1050 - 0402 = 0648$. It is also the method of converting a signed-digit number to a conventional radix number.

1.5 FLOATING-POINT NUMBER REPRESENTATION

All the number representations discussed so far have the radix point in a fixed position, thus they belong to the fixed-point number representation. In computer systems, the radix point is not actually shown, but its existence and position are mutually agreeable by humans and computers.

The position of the radix point determines the number of digits in the integer part and that in the fraction part. Given the total number of digits fixed, the more digits in the integer the bigger number represented. On the other hand, the more digits in the fraction the better precision obtained. A new idea about floating-point number representation is hereby introduced in contrast to the fixed-point number representation. The general format of floating-point representation is

$$F = (M, E) \triangleq M \times r^E.$$

Here, M represents the *mantissa* (or *significand*), and E the *exponent*. r is the radix as usual.

For example, $(-0.0025, +3)$ means -0.0025×10^3 for $r = 10$ and is equal to 2.5_{10}.

The mantissa M and exponent E are both signed numbers. M is usually a signed fraction and E a signed integer. Many ways exist to represent the sign, the fraction and the integer. All computers had different representations before the IEEE standard was published. Here we focus on the most common cases, and other representations are left for readers to analyze.

1.5.1 Normalization

Most often M is a *normalized* fraction in the sign-magnitude form. The true magnitude of it is $|M| = (0.m_{-1}m_{-2} \cdots m_{-k})$. *that is,* $1 \leq |M| < 2$.) By *normalized*, we mean that the MSD of the mantissa equals non-zero, that is, $m_{-1} \neq 0$. If it is zero, the floating-point number is *unnormalized*.

$$\begin{aligned} F &= (-0.0025, +3) \\ &= (-0.25, +1) \end{aligned}$$

indicates how the *normalization* procedure can be conducted. Note that a normalized mantissa has its absolute value limited by

$$\frac{1}{r} \leq |M| < 1,$$

where $\frac{1}{r}$ is the weight carried by the MSD in the mantissa. In other words, if $\frac{1}{r} \leq |M| < 1$, the floating-point number is normalized. For the binary case with k bits in the mantissa,

$$\frac{1}{2} \leq |M| \leq 1 - 2^{-k}. \quad (1.6)$$

The reason for normalization is to fully utilize the available bits which are limited in the computer system and are hence precious. Allowing too many leading 0s in the fraction may cause unnecessary truncation of the lower order bits in the mantissa. The bit positions occupied by those leading 0s are wasted, and the accuracy of the number representation is degraded.

The normalization can be easily done. Just perform left shift on the unnormalized mantissa until the leading 0s are all shifted out and the first nonzero digit reaches the most significant position. In the meantime, adjust the exponent accordingly. Left shifting the mantissa for k positions will enlarge the represented number r^k times, and deducting k from the exponent can reduce the number r^k times and keep the represented number unchanged.

After discussing mantissa M in the floating-point number representation, let's look at the exponent E. Most commonly E is an r's complement integer in the "biased" form, though some computer systems left it "unbiased".

Let E be a 2's complement integer for illustration. The basic principles below are applicable to all the $r > 2$.

1.5.2 Bias

Suppose

$$E = (a_{q-1} a_{q-2} \cdots a_0)_2,$$

where q is the number of bits. Then,

$$-2^{q-1} \leq E \leq 2^{q-1} - 1. \quad (1.7)$$

For example, given $q = 4$,

$$-2^3 \leq E \leq 2^3 - 1.$$

That is,

$$-8 \leq E \leq +7,$$

and
$$(1000)_2 \leq E \leq (0111)_2.$$

Choose a constant as a bias which is usually the magnitude of the lowest bound value of unbiased exponent, *that is,*
$$bias = 2^{q-1}.$$

Add the bias to the unbiased exponent, we can have E_{biased} in the following range:
$$0 \leq E_{biased} \leq 2^q - 1.$$

Recall that $E_{unbiased}$ can be either negative or positive, while E_{biased} are all positive now.

The relationship of $E_{unbiased}$ and E_{biased} can be expressed as follows:
$$E_{unbiased} = E_{biased} - 2^{q-1}.$$

The length of the mantissa determines the precision of the represented number. The range of the number that can be represented, however, is dependent on the length of the exponent. Given a fixed number of bits for the floating-point representation, there are tradeoffs in reserving the number of bits for the mantissa and the number of bits for the exponent.

Consider the binary representation of a floating-point number with a total of 32 bits in a computer system. The radix is in common, and is mutually understood by the human and the computer, hence it is not necessary to be represented. The sign takes one bit position, S, to represent it. If $S = 0$, the sign is positive. If $S = 1$, the sign is negative. The rest of the things we need to represent are the magnitude of the mantissa and the exponent only. Partition the 32 bits into two parts which are reserved for the mantissa and exponent respectively. As a good compromise of the precision and the range of the represented number, 23 bits (from bit 0 to bit 22) are for the mantissa which is represented in the true magnitude form, and 8 bits (from bit 23 to bit 30) are reserved for the exponent which is represented in the biased form. Bit 31 is reserved for sign. Figure 1.1(a) shows the bit partition. Let number $F = -0.6259765 \times 2^{-3}$. The floating-point representation of it is indicated in Figure 1.1(b). Here, the binary sign-magnitude form for the mantissa is $M = (-0.6259765) = (1\ 10100000010000000000000)_2$, the biased exponent is $E_{biased} = -3 + 2^{8-1} = (11111101)_2 + (10000000)_2 = (01111101)_2$.

Given a number F represented in Figure 1.1(a), one should realize that
$$F = (-1)^S |M| \cdot r^{E-bias}. \tag{1.8}$$

Here, S has one of the two values, 0 or 1. If $S = 0$, $(-1)^0 = +1$, meaning that the sign of the represented number is positive. If $S = 1$, $(-1)^1 = -1$, meaning that the

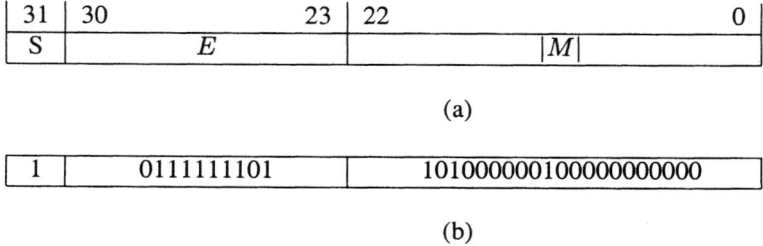

Fig. 1.1: Floating-Point Representation

sign of the represented number is negative. E in the representation is a biased exponent (which is implied hereafter). In the original number, the true exponent should be $E_{unbiased}$.

The reason to utilize the biased exponent is that exponent comparison can be made easier this way. Remember that the comparison of two signed numbers are nontrivial, while the bigger negative number has a smaller magnitude compared with a negative number, and both are in the complement representation.

We discuss below the range and precision of the floating-point number representation.

Let $|M|_{min}$ be the smallest magnitude of the mantissa, and E_{min} the smallest true exponent. ($|M|_{max}$ and E_{max} are the largest mantissa and true exponent, respectively). $|M|_{min} \cdot r^{E_{min}}$ is the smallest positive number that can be represented in the floating-point number system. ($|M|_{max} \cdot r^{E_{max}}$ is the largest positive number that can be represented). Recalling that $\frac{1}{2} \leq |M| < 1$ (Equation (1.6)) and $-2^{q-1} \leq E_{unbiased} \leq 2^{q-1} - 1$ for q binary bits in the exponent (Equation (1.7)), we have the following range for F^+, the positive number that can be represented in the floating-point number system.

$$\frac{1}{2} \cdot r^{-2^{q-1}} \leq F^+ < 1 \cdot r^{2^{q-1}-1},$$

or

$$\frac{1}{2} \cdot r^{-2^{q-1}} \leq F^+ \leq (1 - 2^{-k}) \cdot r^{2^{q-1}-1},$$

where k is the number of bits in the mantissa. Symmetrically, if F^- denotes the negative number that can be represented in the floating-point number system,

$$-(1 - 2^{-k}) \cdot r^{2^{q-1}-1} \leq F^- \leq -\frac{1}{2} \cdot r^{-2^{q-1}}.$$

We can see that F^+_{min} and F^-_{max} are not adjacent to 0. A special case is made to represent value 0 by letting $(M, E) = (0, 0)$. The sub-range of F^- and F^+ are apart

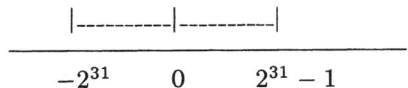

(a) Fixed-Point Numbers

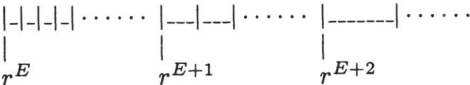

(b) Floating-Point Numbers

Fig. 1.2: Range of the Numbers

Fig. 1.3: Precision of Floating-Point Numbers

from each other. The region in between, except 0, is referred to as underflow.

Given a total of 32 binary bits partitioned as previously described, the range of the integer numbers in the fixed-point number representation is shown in Figure 1.2(a), and the range of the floating-point numbers represented is shown in Figure 1.2(b).

With k bits in the mantissa, the weight carried by the LSB is 2^{-k}. Every single change in the mantissa will be scaled by at least $r^{-2^{q-1}}$ where the exponent is chosen to be $E_{min} = -2^{q-1}$. So the least significant change in the mantissa followed by a minimum scaling results in the closest distance of two consecutive numbers in the floating-point number system, hence determines the precision of the number representation. Note that such precision can be only guaranteed when the exponent is kept as E_{min}. For a bigger exponent, the scale will be enlarged. For $q = 8$ and $k = 23$, Figure 1.3 drafts the distribution of consecutive numbers.

Base r above is machine dependant. DEC/VAX and Cyber 70 use $r = 2$, while IBM/370 has $r = 16$.

Fig. 1.4: Double Precision Floating-Point Representation

The IEEE standard was formulated in 1981 when all different representations existed. The suggested standard unifies all the number system design issues and makes the transfer of data and programs between one computer and another easier.

The IEEE single precision format is similar to the one presented in Figure 1.1 except in the following protocols. The double precision representation is shown in Figure 1.4. What follows is based on the single precision model, while the same principles apply in the double precision case.

Instead of $\frac{1}{r}$ to 1, the range of the mantissa magnitude is 1 to 2 now. Equation (1.8) $F = (-1)^S |M| \cdot r^{E-bias}$, with $|M| = (0.m_{-1}m_{-2}\cdots m_{-k})$, is now

$$F = (-1)^S(1.m_{-1}m_{-2}\cdots m_{-k}) \cdot r^{E-bias}.$$

The same place used to store $|M|$ is now storing f. The mantissa represented is no longer a pure fraction equal to $\pm|M|$. It has an integer part 1 now, and $\pm 1.f$ is the mantissa which can be viewed as a left shifted normalized fraction. Hence,

$$F = (-1)^s(1.f) \cdot r^{e-bias}.$$

Only the fraction part of mantissa, f, is to be recorded, but not the integer 1. 1 is always there, as understood by humans and computers. s is the sign and e is the biased exponent as usual. Another difference here lies in the bias. Rather than $bias = 2^{q-1}$ in the IEEE standard,

$$bias = 2^{q-1} - 1 = 127$$

for $q = 8$.

While $-128 \leq e_{unbiased} \leq 127$,

$$1 \leq e_{biased} \leq 254. \qquad (1.9)$$

Note that adding the bias to the lower bound of $e_{unbiased}$ results in -1 which is not a legal biased number. Besides, $e_{biased} = 0$ is reserved for some special cases. $e_{biased} = 255$ is also used in special representations, as shown in Table 1.3.

Here NAN means *Not A Number*, and is used to represent the result of $X/0$, $\sqrt{-1}$, or so forth, in case such operations are encountered. The representation of gradual underflow is specially designed in which $f \neq 0$ records the fraction part of a mantissa magnitude $0.f$, and $e = 0$ is defined as $e_{unbiased} = -126$. That means if

Table 1.3: Reserved Representation in IEEE Standard

e	f	Representing
0	0	0
0	$\neq 0$	gradual underflow
255	0	$\pm\infty$
255	$\neq 0$	NAN

$e < e_{min} = 1$ (see Equation (1.9)), we enlarge it to the same as e_{min}, and shift the mantissa right at the same time. Hence, the mantissa has the integer part equal to 0 (referred to as "denormal"). Further, the most significant bit(s) in the fraction equals 0(s). When more and more nonzero bits are shifted out and zero bits are shifted in, the represented number is flushed to 0. So, such representation as

$$(-1)^s (0.f) \cdot 2^{-126}$$

is of "gradual underflow".

To find the range of the numbers representable by IEEE single precision form, we have

$$\begin{aligned} F_{min}^+ &= 1.f_{min} \cdot 2^{e_{min}-bias} \\ &= 1.0 \times 2^{1-127} \\ &= 2^{-126}, \end{aligned}$$

recalling that $1 \leq f < 2$ or $1 \leq f \leq 2 - 2^{-23}$, and $1 \leq e_{biased} \leq 254$ by (1.9). On the other hand,

$$\begin{aligned} F_{max}^+ &= 1.f_{max} \cdot 2^{e_{max}-bias} \\ &= (2 - 2^{-23}) \times 2^{254-127} \\ &= 2^{128} - 2^{104}. \end{aligned}$$

The number representations learned so far are as follows.

$$\left.\begin{array}{l} \text{Unsigned Number} \\ \quad \text{Integer} \\ \quad \text{Fraction} \\ \text{Signed Number} \\ \quad \text{Sign-Magnitude} \\ \quad (r-1)'\text{s comp.} \\ \quad r'\text{s comp.} \end{array}\right\} \begin{array}{l} \text{Conventional Radix} \\ \\ \text{Signed Digit} \end{array} \left.\begin{array}{l} \text{Fixed-Point} \\ \text{Floating-Point} \end{array}\right\}$$

Other than these number representations, two more number systems will be introduced below.

1.6 RESIDUE NUMBER SYSTEM

In the Residue Number System (RNS), a set of moduli are given which are independent of each other. An integer is represented by the residue of each modulus and the arithmetic operations are based on the residues individually.

Let $\{m_1, m_2, \cdots, m_n\}$ be a set of positive integers all greater than 1. m_i is called a modulus, and the n-tuple set $\{m_1, m_2, \cdots, m_n\}$ is called a moduli set. Consider an integer number X. For each modulus in $\{m_1, m_2, \cdots, m_n\}$, we have $x_i = X$ mod m_i (denoted as $|X|_{m_i}$). Thus a number X in RNS can be represented as

$$X = (x_1, x_2, \cdots, x_n).$$

For example, let the moduli set $(m_1, m_2, m_3) = (2, 3, 5)$. To represent the number 9 in RNS, we have $x_1 = |X|_{m_1} = |9|_2 = 1$, $x_2 = |X|_{m_2} = |9|_3 = 0$, and $x_3 = |X|_{m_3} = |9|_5 = 4$. So, the RNS representation of 9 is (1,0,4).

Given a specific moduli set $\{m_1, m_2, \cdots, m_n\}$. In order to avoid redundancy, the moduli of a Residue Number System must be pairwise relatively prime. That means they don't have common divisor other than 1.

Let $M = \prod_{i=1}^{n} m_i$. It has been proved that if $0 \leq X < M$, the number X is one-to-one corresponding to the RNS representation.

The above definition can be extended to represent a negative integer in RNS. In addition to the m_is defined as positive integers, x_is are limited to positive integers as well.

For example, given a negative number -9 and the same moduli set as above, we have $x_1 = |-9|_2 = 1$, $x_2 = |-9|_3 = 0$, and $x_3 = |-9|_5 = 1$. So the RNS representation of -9 is (1,0,1). Note, as defined, moduli are positive numbers. Thus in $(-9) \div 2$, instead of having quotient -4 and remainder -1, we let the quotient be $-4 - 1 = -5$, hence the remainder becomes $(-9) - (-5) \times 2 = 1$. Similarly in $(-9) \div 5$, the quotient is $-1 - 1 = -2$, and the remainder should be $(-9) - (-2) \times 5 = 1$.

In general, if X is a signed integer to be represented, the range of X having the one-to-one corresponding RNS representation is as follows. If X is even, then $-\frac{M}{2} \leq X \leq \frac{M}{2} - 1$ with 0 in between the positive and negative sub-ranges. If X is odd, then $-\frac{M-1}{2} \leq X \leq \frac{M-1}{2}$.

The arithmetic operations based on RNS can be performed on different moduli independently to avoid the carry in addition, subtraction and multiplication, which is usually time consuming. However, the comparison and division are more complicated and the fraction number computation is immatured. Due to this, RNS is not yet popular in general-purpose computers, though it is very promising in special-purpose applications.

1.7 LOGARITHMIC NUMBER SYSTEM

Given an unsigned number X we take the logarithm of X based on r, and denote it as L_x, that is,

$$L_x = log_r X.$$

Assume that L_x is represented in binary with n bits in integer and k bits in fraction. Then we have X represented in the Logarithmic Number System (LNS) form by

$$L_x = x_{n-1}x_{n-2}\cdots x_0.x_{-1}x_{-2}\cdots x_{-k}.$$

If $X > 1$, L_x is positive. If $X < 1$, L_x is a negative number. Here the 2's complement representation, for example, applies to express L_x, and L_x denotes a number $X = r^{L_x}$.

For example, given $L_x = (100.1)_2$ and $r = 4$, we have

$$L_x = -3.5$$

representing

$$\begin{aligned} X &= 4^{-3.5} \\ &= \frac{1}{4^3 \times \sqrt{4}} \\ &= \frac{1}{128}. \end{aligned}$$

Note that X takes at least 8 bits to represent in the conventional radix number system, but needs only 4 bits here in the LNS representation.

Alternatively, a biased number can be applied to express L_x, where

$$L_x = x_{n-1}x_{n-2}\cdots x_0.x_{-1}x_{-2}\cdots x_{-k} + bias.$$

The bias is 2^{n-1}, or $2^{n-1} - 1$, and in this case L_x denotes a number $X = r^{L_x - bias}$ in the LNS.

We have discussed the signed number L_x representing an unsigned number X greater than or smaller than 1. Now if X is a signed number, then an additional digit is needed to represent the sign of X. Let S_x be the sign digit. $S_x = 0$ if $X > 0$, and $S_x = 1$ if $X < 0$. Also, L_x is $log_r |X|$ now. The logarithmic number representation for a signed number X is $S_x L_x$, and the value it represents is $X = (-1)^{S_x} r^{L_x}$ (or

$X = (-1)^{S_x} r^{L_x - bias}$ for a biased L_x).

In discussing the range of X that can be represented in the LNS, 0 is excluded since $r^{L_x} \neq 0$ for any L_x. With n bits in the integer part and k bits in the fraction part of L_x, we have the range of unbiased L_x as follows:

$$-2^{n-1} \leq L_x \leq 2^{n-1} - 2^{-k}.$$

So, the positive sub-range of X that can be represented is

$$r^{L_x^{min}} \leq X^+ \leq r^{L_x^{max}},$$

which is

$$r^{-2^{n-1}} \leq X^+ \leq r^{2^{n-1} - 2^{-k}}.$$

The negative sub-range of X that can be represented is

$$-r^{L_x^{max}} \leq X^- \leq -r^{L_x^{min}},$$

which is

$$-r^{2^{n-1} - 2^{-k}} \leq X^- \leq -r^{-2^{n-1}}.$$

Once numbers are represented by their exponents, the multiplication and division of them can be completed by addition and subtraction, respectively. The power and root of them can be obtained by multiplication and division, respectively. However, the addition and subtraction of the LNS numbers are not easy to perform. Also, the logarithm and antilogarithm needed for number conversion are very complex. So the LNS is not widely adopted in general purpose computations, but is very attractive in application oriented arithmetic design.

REFERENCES

1. A. Avizienis, "A Study of Redundant Number Representations for Parallel Digital Computers," Ph.D. Dissertation, University of Illinois at Urbana-Champaign, Digital Computer Laboratory, May 1960.

2. A. Avizienis, "Signed-Digit Number Representations for Fast Parallel Arithmetic," *IRE Trans. Elec. Comp.*, Vol. EC-10 (Sept. 1961), pp. 389–400.

3. A. Avizienis, "Digital Computer Arithmetic: A Unified Algorithmic Specification," *Proc. Symposium on Computers and Automata*, Polytechnic Institute of Brooklyn, 1971, pp. 509–525.

4. J. J. F. Cavanagh, *Digital Computer Arithmetic: Design and Implementation,* McGraw-Hill, New York, 1984.

5. J. F. Couleur, "BIDEC-a Binary-to-Decimal or Decimal-to-Binary Converter," *IRE Trans. Elec. Comp.,* vol. EC-7, no. 4, Dec. 1958, pp. 313–316.

6. I. Flores, *The Logic of Computer Arithmetic,* Prentice-Hall, Englewood Cliffs, NJ, 1963.

7. H. L. Garner, "Number Systems and Arithmetic," in *Advances in Computers,* Vol. 6, F. L. Alt and M. Rubinoff, eds. Academic Press, New York, 1965. pp. 131–194.

8. H. L. Garner, "A Survey of Recent Contributions to Computer Arithmetic," *IEEE Trans. Comp.,* Vol. C-25, No. 12, Dec. 1976, pp. 1277–1282.

9. D. Goldberg, "Computer Arithmetic," in *Computer Architecture: A Quantitative Approach,* D. A. Patterson, and J. L. Hennessy, Morgan Kaufmann, San Mateo, CA, 1996.

10. J. B. Gosling, *Design of Arithmetic Units for Digital Computers,* Springer-Verlag, New York, 1980.

11. K. Hwang, *Computer Arithmetic: Principles, Architecture and Design,* Wiley, New York, 1978.

12. K. Hwang and T. P. Chang, "A New Interleaved Rational/Radix Number System for High-Precision Arithmetic Computations," *Proc. Fourth Symposium on Computer Arithmetic,* Oct. 1978.

13. I. Koren and Y. Maliniak, "On Classes of Positive, Negative and Imaginary Radix Number Systems," *IEEE Trans. Comp.,* C-30 (May 1981) pp. 312–317.

14. D. E. Knuth, *The Art of Computer Programming,* Vol. 2, *Seminumerical Algorithms,* Addison-Wesley, Reading, MA, 1969, Chap. 4.

15. U. Kulisch, "Mathematical Foundations of Computer Arithmetic," *IEEE Trans. Comp.,* Vol. C-26, No. 7, July 1977, pp. 610–620.

16. W. C. Lanning, "Automata for Direct Radix Conversion," in *Computers and Electrical Engineering,* vol. 1, Pergamon Press, Elmsford, NY, 1973, p. 281.

17. J. D. Marasa, "Accumulated Arithmetic Error in Floating-Point and Alternative Logarithmic Number Systems," M.S. Thesis, Seven Institute Technology Washington University, St. Louis, MO, June 1970.

18. D. W. Matula, "Number Theoretic Foundations for Finite-Precision Arithmetic," in *Applications of Numbers Theory to Numerical Analysis,* W. Zaremba ed. Academic Press, New York, 1972, pp. 479–489.

19. D. W. Matula, "Fixed-Slash and Floating-Slash Rational Arithmetic," *Proc. Third Symposium on Comp. Arith.*, IEEE Catalog No. 75 CH1017-3C, Nov. 1975, pp. 90–91.

20. D. W. Matula, "Radix Arithmetic: Digital Algorithms for Computer Architecture," in *Applied Computation Theory: Analysis, Design, Modeling*, R.T. Yeh ed., Prentice-Hall, Englewood Cliffs, NJ, 1976, Chap. 9.

21. J.-D. Nicoud, "Iterative Arrays for Radix Conversions," *IEEE Trans. Comp.*, vol. C-20, No. 12, Dec. 1971, pp. 1479–1489.

22. B. Parhami, "Generalized Signed-Digit Number Systems: A Unifying Framework for Redundant Number Representations," *IEEE Trans. Comp.*, vol. 39, Jan. 1990, pp. 89–98.

23. J. E. Robertson, "Redundant Number Systems for Digital Computer Arithmetic," Notes for the Univ. of Michigan Engineering Summer Conference, in "Topics in the Design of Digital Computing Machines," Ann Arbor, MI, July 6–10, 1959.

24. N. R. Scott, *Computer Number Systems and Arithmetic*, Prentice-Hall, Englewood Cliffs, NJ, 1985.

25. O. Spaniol, *Computer Arithmetic: Logic and Design*, Wiley, New York, 1981.

26. E. E. Swartzlander, Jr., ed., *Computer Arithmetic*, Vol. 1, IEEE Computer Society Press, Los Alamitos, CA, 1990.

27. E. E. Swartzlander, Jr., ed., *Computer Arithmetic*, Vol. 2, IEEE Computer Society Press, Los Alamitos, CA, 1990.

28. N. S. Szabo and R. I. Tanaka, *Residue Arithmetic and Its Applications to Computer Technology*, McGraw-Hill, NY, 1967.

29. C. Tung, "Arithmetic," *in Computer Science*, A. F. Cardenas et al., ed., Wiley-Interscience, New York, 1972, Chap. 3.

30. S. Waser and M. J. Flynn, "Introduction to Arithmetic for Digital System Designers," Holt, Rinehart, Winston, New York, 1982.

PROBLEMS

1.1 Given $X = 237.4$, what decimal value does it represent if
(a) It is an octal number?
(b) It is a hexadecimal number?

1.2 Let n be the number of integer digits and k the number of fraction digits. Given a decimal number $Y = 19.125$,

(a) Represent Y in binary with $(n, k) = (5, 3)$;
(b) Represent Y in octal with $(n, k) = (2, 1)$.

1.3 (a) Convert $(1101111.00101)_2$ to a hexadecimal number.
(b) Convert $2A3C.18_{16}$ to an octal number.

1.4 Let $X = (x_{n-1} \cdots x_1 x_0)_2$. Verify
(a) The sum of X and the 1's complement of X is 0;
(b) The sum of X and the 2's complement of X is -0.
From the above we can see that $X + (-X) = 0$ where $-X$ can be represented in 1's or 2's complement form.

1.5 Fill the following table if radix $r = 2$, the number of bits $n = 6$ (including the sign bit), and the absolute magnitude $|A| = 11_{10}$.

	$A >= 0$	$A < 0$
Sign-magnitude form		
Diminished radix complement form		
Radix complement form		

1.6 Find the diminished radix complement number and radix complement number of
(a) -572_8;
(b) -0.456_8.

1.7 Given a digit sequence 7052, what value does it represent if it is in:
(a) Unsigned octal number system?
(b) Unsigned hexadecimal number system?
(c) Signed octal number system with diminished radix complement representation?
(d) Signed octal number system with radix complement representation?

1.8 (a) Convert 725_8 to a signed-digit number with $n = 4$ and $\alpha = 5$.
(b) Convert -725_8 to a signed-digit number with $n = 4$ and $\alpha = 5$.

1.9 Given the following bit format, represent 0.375 and -104 as normalized floating-point numbers (not necessarily biased).

1.10 Given a floating-point number $F = (m, e) = (0011, 0101)$,
(a) if m is a binary sign-magnitude fraction and e is an unbiased 2's complement

12	6	5	0
m		e	

exponent, $F = $ _____;
(b) as in (a) but e is a biased exponent with a bias constant 2^{q-1} where q is the number of bits of the exponent, $F = $ _____.

1.11 MIPS computer system adopts the single precision IEEE format in its floating point number representation, that is, (a) Represent $(-0.625)_{10}$ in the MIPS system.

1	8 bits	23 bits
s	e	f

(b) Given the following number in the MIPS floating point form, what does it represent?

1	8 bits	23 bits
0	10000100	00110000000000000000000

1.12 (a) With the IEEE single precision floating point form, what are the decimal values that can be represented?
(b) With the IEEE double precision floating point form, what are the binary values that can be represented?

2
Addition and Subtraction

Addition and subtraction are the basic operations in computer arithmetic. Multiplication and division are based on addition and subtraction. Fast adders (subtractors) are desirable not only for speeding up fundamental operations like addition and subtraction in arithmetic operation, but also for accelerating the multiplication and division which involve massive addition and subtraction. We discuss in this chapter the variation of two-operand adders with interest in their time complexity and area complexity. Ripple carry adders, conditional-sum adders, carry-completion adders and carry-lookahead adders will be introduced.

2.1 SINGLE-BIT ADDERS

2.1.1 Logical Devices

The Arithmetic Logic Unit (ALU) design aims to minimize time complexity for achieving high speed, as well as area complexity for cost reduction. In this book we limit our discussion to the logic level design. With a wide variety of logic families rapidly advancing, the electronic implementation is left for further study with trade offs in the density and cost and the speed and power dissipation.

Denote the time complexity as $\triangle_T$, and the area complexity as A_T. We list in Table 2.1 the $\triangle_T$ and A_T for various logic gates which serve as fundamental units in ALU design. We limit the logic gates to n-input gates where $1 \leq n \leq 10$. The measure of $\triangle_T$ is $1\triangle_g$ which is the propagation delay of a NAND gate or a NOR

Table 2.1: Delay Time and Area of Logic Gates

NAND	$1\triangle_g$	$1A_g$
NOR	$1\triangle_g$	$1A_g$
NOT	$1\triangle_g$	$1A_g$
AND	$2\triangle_g$	$2A_g$
OR	$2\triangle_g$	$2A_g$
XOR	$2\triangle_g$	$3A_g$
XNOR	$2\triangle_g$	$3A_g$
AOI	$1\triangle_g$	$2A_g$

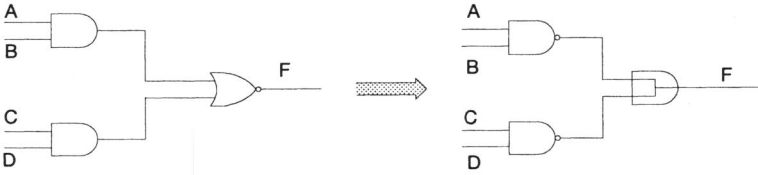

Fig. 2.1: AOI Function

gate. The measure of A_T is A_g which is the area required by one NAND gate or NOR gate.

An AND-OR-Invert (AOI) function can be realized by hard-wired circuit wiring the high-impedance outputs of two NAND gates together (see Figure 2.1(a)). Actually, $F = \overline{AB} \cdot \overline{CD} = \overline{AB + CD}$ here. For a limited RC (resistance and capacitance) load, the $\triangle_T$ is $1\triangle_g$, and the A_T is $2A_g$ for two inputs.

An n-to-2^n Decoder has n input lines called address lines and 2^n output lines. 2^n AND gates are available, each providing an output line and each having n inputs connected to the variables on the address lines or the complement of them. According to the combination of the logic values of the address lines, one and only one of the 2^n output lines should go "high."

Figure 2.2(a) is a schematic of a 2-to-4 decoder with an enable line. The delay time of the Decoder is $3\triangle_g$ and the area complexity is $10\triangle_g$.

A multiplexer (MUX) is a many-to-one function unit which has many inputs and one output. Precisely, a 2^n-to-1 multiplexer has n control input lines and 2^n data input lines. It is based on an n-to-2^n decoder. Each of the 2^n AND gates is given a data input. The address lines of the decoder are used to select the gate. Only the selected gate can route the data to the output.

Figure 2.2(b) shows a 4-to-1 MUX based on a 2-to-4 decoder to select one data out of four and transmit it to the output.

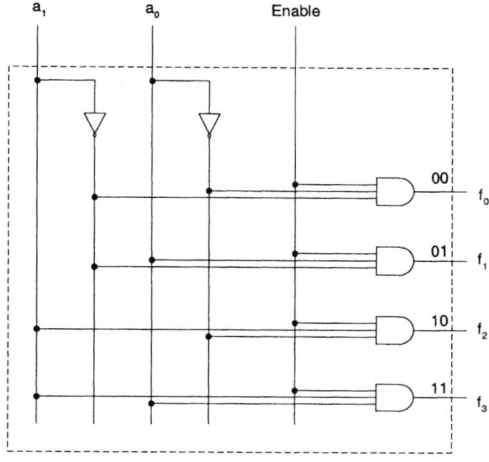

(a) 2-to-4 decoder

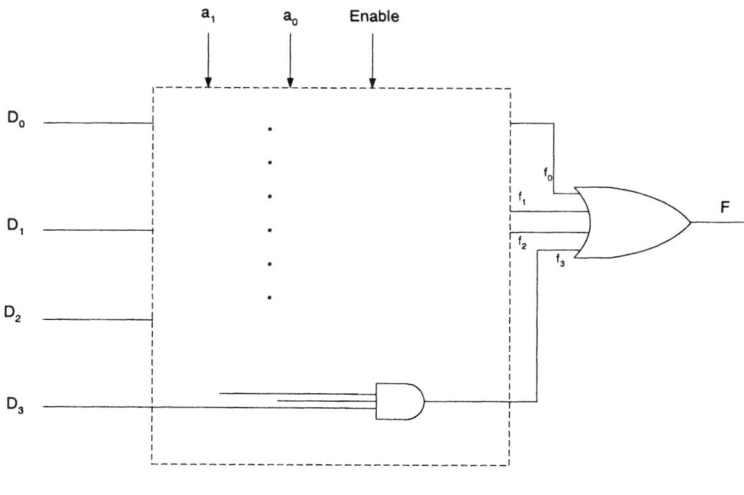

(b) 4-to-1 Multiplexer

Fig. 2.2: Decoder and Multiplexer

Table 2.2: Logic Function of a Half-Adder

a_i	b_i	c_{i+1}	s_i
0	0	0	0
0	1	0	1
1	0	0	1
1	1	1	0

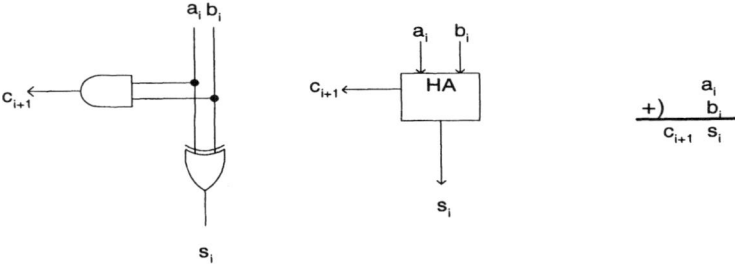

Fig. 2.3: Single-Bit Half-Adder

The hardware can be implemented by a wired AOI and an invertor. Hence, the delay time of the 4-to-1 MUX is $3\triangle_g$, and the area complexity is $7A_g$.

Quad 2-to-1 MUXs, dual 4-to-1 MUXs, dual 8-to-1 MUXs and single 16-to-1 MUX are available on IC (Integrated Circuit) chips currently.

2.1.2 Single-Bit Half-Adder and Full-Adders

A single-bit half-adder(HA) can perform the addition of two bits. Let the two input bits be a_i and b_i. The output consists of two bits, a sum bit s_i and a carry-out bit c_{i+1}. The logic function of the HA to be performed is listed in Table 2.2.

We can see that the logic expression of the sum bit is

$$s_i = a_i \oplus b_i,$$

and that of the carry-out bit is

$$c_{i+1} = a_i b_i.$$

Figure 2.3 shows the logic symbol and schematic of a half-adder.

Just like c_{i+1} which is generated in bit position i and carried to bit position $i + 1$, each bit position i will receive a carry generated in the adjacent lower order bit position. We refer to it as c_i, the carry-in of bit position i. So, besides the two input bits a_i and b_i, a third input bit c_i should also be considered. An adder which can add three bits together is called a full-adder(FA). The logic function of a single-bit FA is listed in Table 2.3.

Table 2.3: Logic Function of a Full-Adder

a_i	b_i	c_i	c_{i+1}	s_i
0	0	0	0	0
0	0	1	0	1
0	1	0	0	1
0	1	1	1	0
1	0	0	0	1
1	0	1	1	0
1	1	0	1	0
1	1	1	1	1

The logic expression of the sum bit can be presented as

$$s_i = a_i \oplus b_i \oplus c_i, \tag{2.1}$$

and the carry-out bit can be found by

$$c_{i+1} = a_i b_i + b_i c_i + a_i c_i. \tag{2.2}$$

Actually, the outputs can be viewed as a two bit number which counts the number of 1s in the input. If only a single 1 exists, $c_{i+1}s_{i+1} = 01$. If there are two 1s, $c_{i+1}s_{i+1} = 10$. If three 1s exist, $c_{i+1}s_{i+1} = 11$.

Figure 2.4 shows the logic symbol and schematic of a single-bit full-adder (FA) with three inputs and two outputs.

$$\begin{aligned} c_{i+1} &= a_i b_i + (a_i \oplus b_i) c_i \\ &= a_i b_i + \bar{a}_i b_i c_i + a_i \bar{b}_i c_i \\ &= a_i b_i + b_i c_i + a_i c_i, \end{aligned}$$

the same c_{i+1} as in Equation (2.2) yields after absorbing $\bar{a}_i$ in the second term and $\bar{b}_i$ in the third term.

After expending Equation (2.1), we have

$$s_i = \bar{A}\bar{B}C + ABC + \bar{A}B\bar{C} + A\bar{B}\bar{C}.$$

In minterm expression $s_i = \{1\} + \{7\} + \{2\} + \{4\}$. That is,

$$\begin{aligned} s_i &= \overline{\{0\} + \{3\} + \{5\} + \{6\}} \\ &= \overline{\bar{A}\bar{B}\bar{C} + \bar{A}BC + A\bar{B}C + AB\bar{C}}, \end{aligned}$$

which can be implemented by the AOI function, and so does c_{i+1}. For such implementation the delay time is

$$\triangle_T = 2\triangle_g,$$

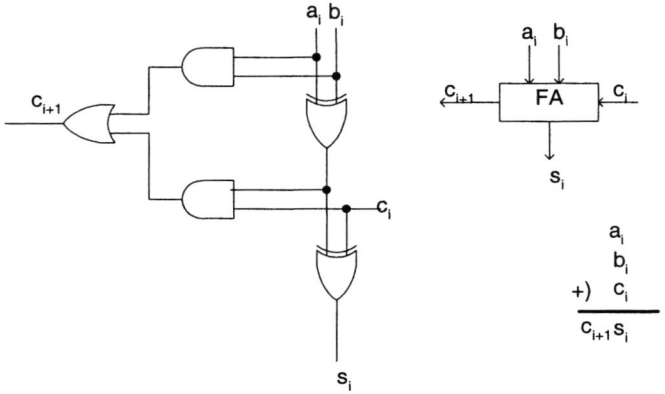

(a) Single-Bit Full-Adder

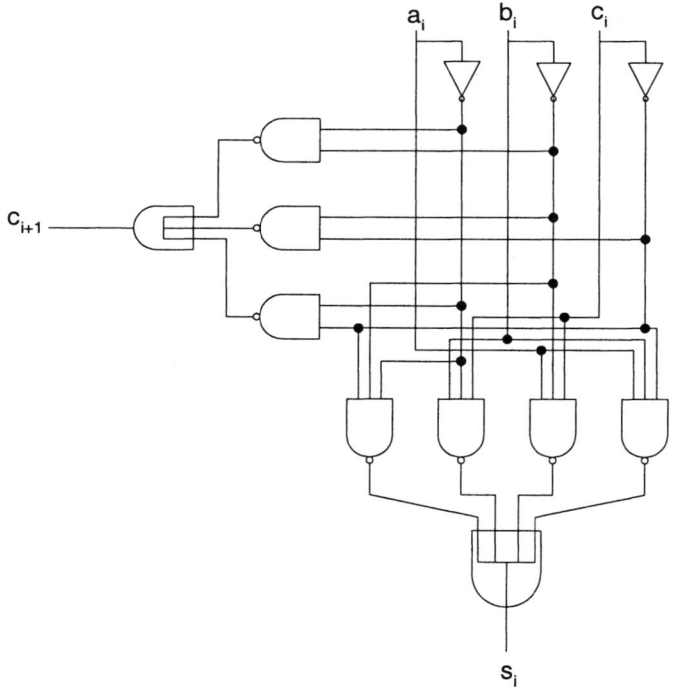

(b) AOI Implemented Full-Adder

Fig. 2.4: Design of Full-Adder

Table 2.4: Single-Bit Subtractor

a_i	b_i	l_i	l_{i+1}	d_i
0	0	0	0	0
0	0	1	1	1
0	1	0	1	1
0	1	1	1	0
1	0	0	0	1
1	0	1	0	0
1	1	0	0	0
1	1	1	1	1

and the area complexity is

$$A_T = 10 A_g.$$

When a single-bit subtraction is performed, say $a_i - b_i$, the deference is represented by bit d_i. Besides, a borrow bit is involved. Let a borrow request made to bit position i be a borrow-in signal denoted as l_i implying a loan from position i. The borrow-out bit l_{i+1} is a borrow request made to bit position $i+1$, or a loan from bit position $i+1$. The truth table of the input and output is listed in Table 2.4.

The logic expression of the difference bit in terms of the three input bits is similar to that of the sum bit in a full-adder. Namely,

$$d_i = a_i \oplus b_i \oplus l_i.$$

The borrow-out bit can be found by

$$l_{i+1} = \bar{a}_i b_i + b_i l_i + \bar{a}_i l_i. \tag{2.3}$$

Figure 2.5 shows a a single-bit subtractor with three inputs and two outputs. l_{i+1} in its sum-of-product realization is

$$\begin{aligned} l_{i+1} &= \bar{a}_i b_i + \overline{a_i \oplus b_i}\, l_i & (2.4) \\ &= \bar{a}_i b_i + a_i b_i l_i + \bar{a}_i \bar{b}_i l_i & (2.5) \\ &= \bar{a}_i b_i + b_i l_i + \bar{a}_i l_i, & (2.6) \end{aligned}$$

the same as in Equation (2.3), after absorbing a_i in the second term and $\bar{b}_i$ in the third term.

2.2 NEGATION

Given a number A, the negation operation finds $-A$, that is the complement number of A. If A is a positive number, after negation we have a negative number. Conversely, if A is a negative number then negation will result in a positive number.

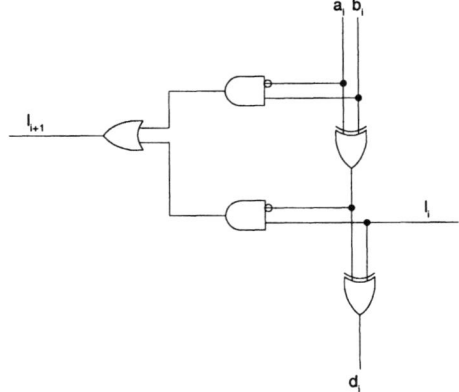

Fig. 2.5: Single-Bit Subtrator

Table 2.5: Negation in One's Complement System

E	a_i	a'
0	0	0
0	1	1
1	0	1
1	1	0

We introduce in this section the negation operation in 1's and 2's complement systems.

2.2.1 Negation in One's Complement System

In the 1's complement system we perform a bit-wise NOT function to obtain a negative number of a given number, that is, change 0 to 1 and change 1 to 0. The circuit shown in Figure 2.6 conducts the negation of an n-bit number in the 1's complement system. In (a), n NOT gates are provided for the simplest negation operation. However, very often the negation function is enabled on-line. For example, when $A + B$ is to be calculated, no negation is necessary. While in the $A - B$ calculation, B is to be negated first and then $A + (-B)$ can be performed.

With a control line, the circuit in (b) can either send the n-bit input number to the output without change, or find the 1's complement of the input number before sending it out. Note that when one input of the exclusive OR gate is set to 1, the other input will be negated before output. The circuit in Figure 2.6(a) and (b) are also referred to as 1's complementer.

The truth table for the enable signal E, the input a_i and the output a'_i is given as follows.

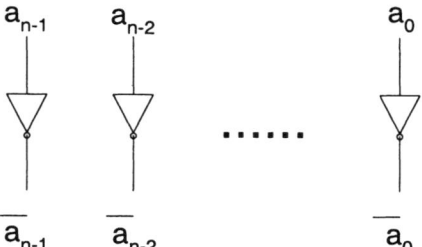

(a) Realizing Negation by NOT Gates

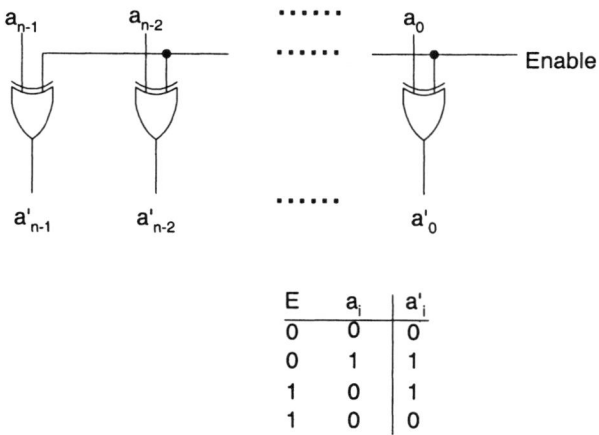

E	a_i	a'_i
0	0	0
0	1	1
1	0	1
1	0	0

(b) Negation by XOR Gates

Fig. 2.6: Negation in One's Complement System

2.2.2 Negation in Two's Complement System

To negate a number in the 2's complement system for a given number

$$a_{n-1} \cdots a_i \cdots a_0,$$

there are two methods to find the 2's complement number of it. The first method in pencil-and-paper description is as follows:

$$+) \quad \frac{\bar{a}_{n-1} \cdots \bar{a}_i \cdots \bar{a}_0}{a^*_{n-1} \cdots a^*_i \cdots a^*_0.}$$

That means to negate an n-bit number in the 2's complement system, one can first find the 1's complement number of it and then increase by 1 using an n−bit half-adder. Note that there is no carry-in but a "1" to add on for the LSB position, while nothing to add on but a carry-in for the rest of the bit positions. The logic circuit is shown in Figure 2.7(a).

The delay time of such an n-bit 2's complementer can be found as

$$\begin{aligned} \triangle_T &= \triangle_{Inverter} + n(\triangle_{AND}) \\ &= (1+2n)\triangle_g. \end{aligned}$$

Here the delay time of the XOR in the half-adder is overlapped with the delay of the AND gate.

The area complexity of this complementer is calculated as follows:

$$\begin{aligned} A_T &= n(A_{Inverter} + A_{AND} + A_{XOR}) \\ &= n(1+2+3)A_g \\ &= 6nA_g. \end{aligned}$$

In the pencil-and-paper manner, the second method to negate a number in the 2's complement system is as follows: Change 0 to 1 and change 1 to 0 from left to right until the right most 1, and then copy that 1 as well as 0s on the right of it if there are any.

Suppose the given number is 10110100, we have

```
1 0 1 1 0 1 0 0
| | | | |
0 1 0 0 1 1 0 0
```

In the above example from left-to-right sub-string 10110 is changed to 01001, and the underlined bit string 100 is copied. Actually this method is consistent with the

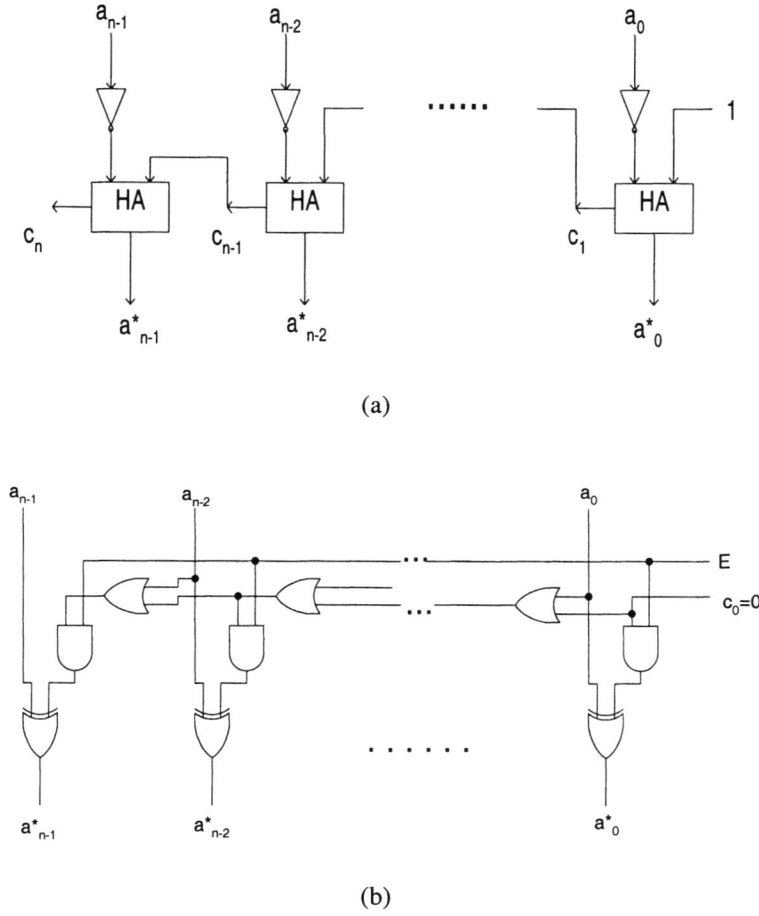

Fig. 2.7: Negation in Two's Complement System

first method. The rightmost 1 (followed by consecutive 0s) becomes the "rightmost 0" (followed by consecutive 1s) after the 1's complement operation. Then in the increment by 1 operation those consecutive 1s can propagate the carry bit by bit, and only that "rightmost 0" can stop the carry propagation. $(01\cdots 1)+1=10\cdots 0$, the same sub-string as given and that is why we copy this part without even performing the increment operation.

Look at the circuit in Figure 2.7(b). When E is set to 1, the chain of OR gates will control which sub-string of input bits is to be altered and which remains the same. c_0 is initialized as 0. If a_0 is 0, the OR gate will propagate a 0 to the left and so is a_1. Wherever the rightmost 1 is hit, a "1" will be propagated to the left. On the right of that "1" where the chain of OR gates outputs a 0, output bit a_i^* will be the same as input bit a_i. On the left of that "1" where the chain of OR gates outputs a "1", output bit a_i^* will be a flipped a_i due to the function of the XOR gate. The above happens when E is set to 1 and the negation in 2's complement system is expected. If E is set to 0, then a_i^* is always the same as a_i and the input number remains unchanged. To compare with the 2's complementer built by the first method, we calculate below the time and area complexity of the 2's complementer built by the second method without the enable line.

$$\begin{aligned}\triangle_T &= (n-1)\triangle_{OR} + \triangle_{XOR} \\ &= [(n-1)\times 2 + 2]\triangle_g \\ &= 2n\triangle_g.\end{aligned}$$

$$\begin{aligned}A_T &= n(A_{OR} + A_{XOR}) - A_{OR} \\ &= [n(2+3) - 2]A_g \\ &= (5n-2)A_g.\end{aligned}$$

2.3 SUBTRACTION THROUGH ADDITION

Let $A = (a_{n-1}\cdots a_1 a_0)$ and $B = (b_{n-1}\cdots b_1 b_0)$ be two signed numbers. $S = A + B$ is also a signed number represented as $(s_{n-1}\cdots s_1 s_0)$. A is referred to as *augend* and B *addend*, or both A and B are called *summand*. S is called the sum of A and B. In the 1's or 2's complement addition, the two input numbers, A and B, are in the 1's or 2's complement representation, and the result S is also represented in the complement form.

In subtraction $D = A - B$, A is referred to as *minuend* and B *subtrahend*. D is the difference of A and B. Subtraction $A - B$ is performed by $A + (-B) = A + \bar{B}$ where $\bar{B}$ is the complement number of B.

We have designed a single-bit full-adder which can be applied to perform the two-operand addition. We are not expected to design a separate, totally new hardware unit to perform the subtraction since in computer arithmetic subtraction is completed

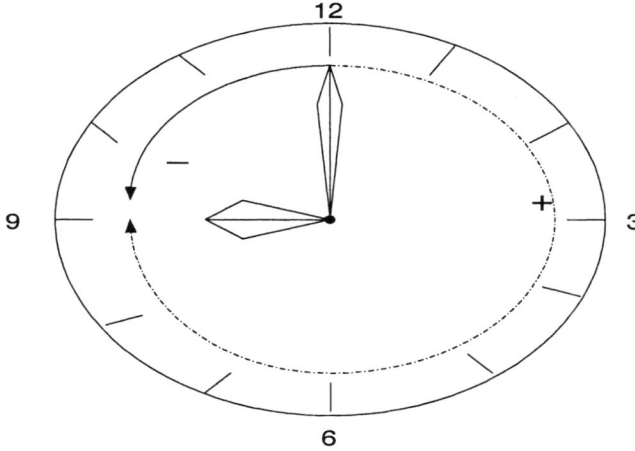

Fig. 2.8: Subtraction through Addition

through addition. The same hardware unit designed for addition can be applied to perform the subtraction.

Consider moving the pointer of a clock. Clockwise move is analogous to addition and counter-clockwise move is analogous to subtraction. For example, suppose a pointer is pointing to 12 o'clock at the beginning. Moving it counter-clockwise for 3 hours will result in 9, that is, $12 - 3 = 9$. On the other hand, moving it clockwise for 9 hours will make it stop at the same place because $(12+9) \bmod 12 = 9$. (See Figure 2.8.) Here 3 and 9 are the complement numbers of each other, the 12's complement numbers. So, $12 - 3$ can be completed by $12 + (-3)$, while (-3) is represented in the complement form, that is, 9.

The same idea applies in computer systems except that the 1's or 2's complement system is adopted rather than the 12's complement system. Given a number B, the negation function and the hardware designed for it can find $(-B)$ as described in Section 2.2.

By combining an adder and a complementor, a unit capable of both addition and subtraction can be designed. In Figure 2.9 a single-bit full-adder has its output

$$c_{i+1} = (a_i + c_i)b_i + a_i c_i \qquad (2.7)$$
$$= a_i b_i + b_i c_i + a_i c_i, \qquad (2.8)$$

which is actually the same logic shown in Equation (2.2). Where b_i used to be at the input of the full-adder, $b_i \oplus M$ is utilized now. Here M is the control signal. When $M = 0$, b_i is input to the adder unchanged and $a_i + b_i$ is performed. When $M = 1$, input to the adder is $\bar{b}_i$ which is the 1's complement number of b_i, and $a_i + (-b_i) = a_i - b_i$ is performed. Hence, Equations (2.1) and (2.7) become

$$s_i = a_i \oplus (b_i \oplus M) \oplus c_i$$

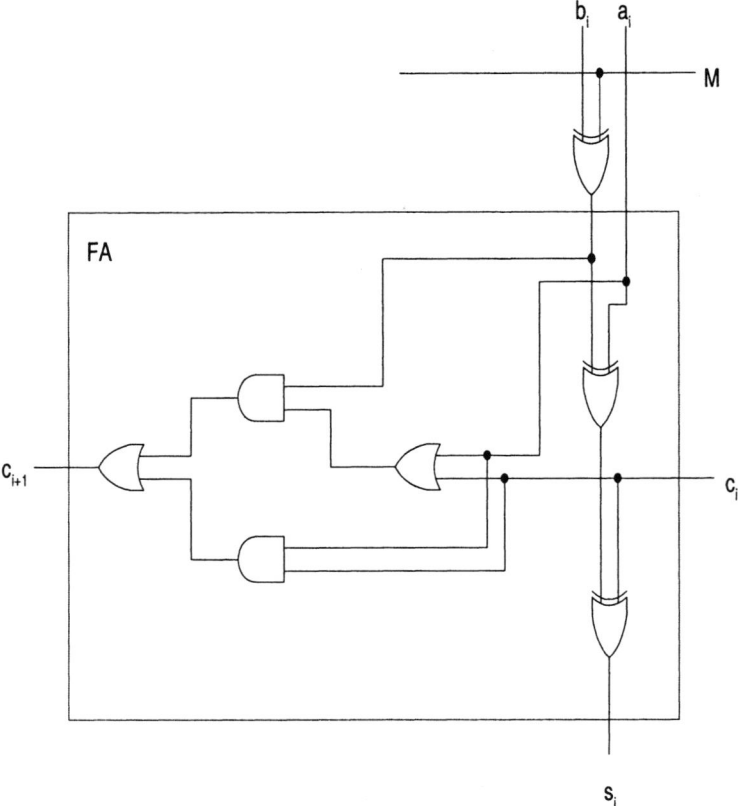

Fig. 2.9: One-Bit Adder/Subtractor

and

$$c_{i+1} = (a_i + c_i)(b_i \oplus M) + a_i c_i,$$

respectively.

To add two n-bit numbers or subtract one from the other, n such units as in Figure 2.9 can be connected in cascade to form an n-bit adder/subtractor. Note that M is the same signal to all the units. Depending on whether M equals to 0 or 1, the negation function is disabled or enabled and an addition or subtraction is to be performed. The mode control is hence

$$M = \begin{cases} 0 & \text{Addition} \\ 1 & \text{Subtraction.} \end{cases}$$

2.4 OVERFLOW

Suppose $A = (a_{n-1} \cdots a_1 a_0)$, $B = (b_{n-1} \cdots b_1 b_0)$ and sum $S = (s_{n-1} \cdots s_1 s_0)$ are the signed numbers and $S = A + B$. The most significant bits (MSB) a_{n-1}, b_{n-1} and s_{n-1} are the sign bits. By $S = A + B$, we mean

$$
\begin{array}{r}
a_{n-1}\ a_{n-2}\ \cdots\ a_0 \\
+)\quad b_{n-1}\ b_{n-2}\ \cdots\ b_0 \\
\hline
s_{n-1}^{\nwarrow c_{n-1}} s_{n-2}^{\nwarrow c_{n-2}} \cdots\ s_0.
\end{array}
$$

In this section, the subscript in the notation for carry is changed as follows. c_{i-1} stands for the carry-in of bit position i, and c_i is the carry-out accordingly.

In what follows, we discuss "overflow" based on the 2's complement system. The scenario holds for other systems too. Depending on the sign of augend and addend, there are three possible cases in the addition.

(1) When both A and B are positive, $a_{n-1} = b_{n-1} = 0$. There should exist $s_{n-1} = 0 + 0 = 0$ meaning the result is a positive number. If $s_{n-1} = 1$, *overflow* occurs. Overflow causes errors in the arithmetic operation – adding two positive numbers cannot result in a negative number. Such error is due to the carry-in bit $c_{n-2} = 1$ generated by the lower order bits. Since the number of the bits left are not sufficient to represent the result of addition, a carry-out entered the sign bit (MSB) position and caused the sign of the result incorrect. That is,

$$\text{if } |A| + |B| < 2^{n-1}, \quad \text{no overflow.}$$
$$\text{if } |A| + |B| \geq 2^{n-1}, \quad \text{overflow occurs.}$$

(2) When both A and B are negative, $a_{n-1} = b_{n-1} = 1$. The correct result should be a negative number with $s_{n-1} = 1$. s_{n-1} will be 1 only if $c_{n-2} = 1$, that is,

$$(2^n - |A|) + (2^n - |B|) \ (mod\ 2^n) \geq 2^{n-1},$$
$$2^n - 2^{n-1} \geq |A| + |B|.$$

Recall that the negative numbers A and B are represented in 2's complement form. Hence,

$$\text{if } |A| + |B| \leq 2^{n-1}, \quad \text{no overflow.}$$
$$\text{if } |A| + |B| > 2^{n-1}, \quad \text{overflow occurs.}$$

(3) We discuss below the case that A is positive and B negative. The case that B is positive and A negative can be viewed in a symmetric way. Since $A = |A|$ and $B = 2^n - |B|$, $A + B = |A| + (2^n - |B|) = 2^n + |A| - |B| \ (mod\ 2^n)$. We have the following:

$$\text{if } |A| < |B|, \quad A + B = 2^n - (|B| - |A|).$$

The result is correct and no overflow occurs.

$$\text{if } |A| \geq |B|, \quad A + B = |A| - |B|.$$

after the *mod* function. No overflow occurs and the correct result is obtained.

To further explore the conditions under which overflow occurs, we investigate the carry-in, c_{n-2}, and carry-out, c_{n-1}, on the MSB position. For case (1) (both A and B are positive) there always exists $c_{n-1} = 0$. To result in a positive sum with $s_{n-1} = 0$, $c_{n-2} = 0$ is expected. For case (2) (both A and B negative) there always exists $c_{n-1} = 1$. To result in a negative sum with $s_{n-1} = 1$, $c_{n-2} = 1$ is expected. For case (3) (A and B have different signs) overflow will not occur. Notice that in this case $c_{n-1} = 0$ if $c_{n-2} = 0$, and $c_{n-1} = 1$ if $c_{n-2} = 1$. Hence, a very easy way to detect overflow is to examine whether c_{n-1} is the same as c_{n-2}. If yes, no overflow. Otherwise, overflow occurs. Hence,

$$Overflow = c_{n-1} \oplus c_{n-2}.$$

2.5 RIPPLE CARRY ADDERS

2.5.1 Two's Complement Addition

Composed of n negation units and n one-bit full adders (FA), a 2's complement adder/subtractor is depicted in Figure 2.10. An overflow detection scheme is also included. Either addition or subtraction can be performed depending on the setting of the mode control M. When $M = 0$, number A and B will be input to the full-adder (FA) and the addition of A and B will be performed. When $M = 1$, the 2's complement number of B will be input to the full-adder and a subtraction will be performed.

Note that the negation unit previously described can only flip the input bit and perform the 1's complement function. For the 2's complement operation, an extra increment is needed. One can see that in the figure, the carry-in of the LSB position c_0 is connected to mode M. When a subtraction is performed $c_0 = M = 1$, and the "plus 1" operation will be carried out together with the addition of the two operands. So added to A are the 1's complement number of B and a "1". That means the 2's complement of B is added to A, which is essentially an $A - B$ operation.

From Figure 2.10 we can see that the carry is rippled to the left across the FAs. The carry-in to bit position i cannot be ready unless the addition in bit position $i - 1$ is completed. This kind of adder is known as a *ripple carry adder*.

The time complexity of the 2's complement adder is

$$\begin{aligned}
\triangle_T &= \triangle_{XOR} + n(\triangle_{FA}) + \triangle_{XOR} \\
&= (2 + 2n + 2)\triangle_g \\
&= (2n + 4)\triangle_g.
\end{aligned}$$

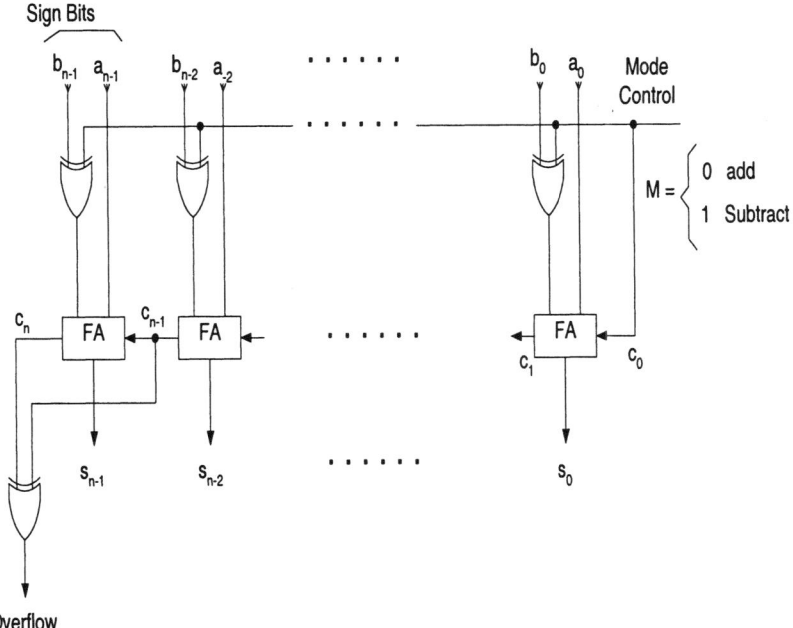

Fig. 2.10: Two's Complement Addition/Subtraction

The area complexity of the 2's complement adder is

$$\begin{aligned} A_T &= n(A_{XOR} + A_{FA}) + A_{XOR} \\ &= (n(3+10)+3)A_g \\ &= (13n+3)A_g. \end{aligned}$$

2.5.2 One's Complement Addition

Given two 1's complement numbers, say A and B, we discuss in this section the design of an adder which can add the two numbers and generate a result, say S, represented in the 1's complement form. Special attention will be paid to a scheme called "end-around carry".

The end-around carry scheme routes the carry-out signal of the MSB position c_n to the LSB position where it is used as a carry-in signal c_0. With the end-around carry, a 2's complement adder described previously can perform a 1's complement addition. We discuss below case by case, the correctness of including such a scheme in a 1's complement adder.

(i) When both A and B are positive, the 1's complement addition can be performed in the same way as the 2's complement addition. The carry-out signal $c_n = 0$ and routing it to LSB will not affect the result of addition.

(ii) When both A and B are negative $c_n = 1$. Comparing the result generated by a circuit without end-around carry to the correct result expected, we will see whether this c_n should be routed to LSB.

Without the end-around carry the circuit performs

$$\begin{array}{rl} & 2^n - 1 - |A| \\ +) & 2^n - 1 - |B| \\ \hline (\text{after } mod) & 2^n - 1 - (|A|+|B|) - 1. \end{array}$$

The expected correct result should be $2^n - 1 - (|A|+|B|)$. A "1" should be added to the result generated by the circuit without end-around carry, and c_n can provide such a "1".

(iii) When A is negative and B is positive (the case of A being positive and B negative can be analyzed in a symmetric way) two possible situations may exist.

a) $|A| > |B|$.

Without the end-around carry, the circuit performs

$$\begin{aligned} & 2^n - 1 \quad -|A| + |B| \\ =\ & 2^n - 1 \quad -(|A| - |B|) \\ <\ & 2^n - 1. \end{aligned}$$

The expected correct result is the same as above. No end-around carry should be asserted. Since the result generated by the circuit is $< 2^n - 1$ and $c_n = 0$, routing c_n to LSB will not affect the result of addition.

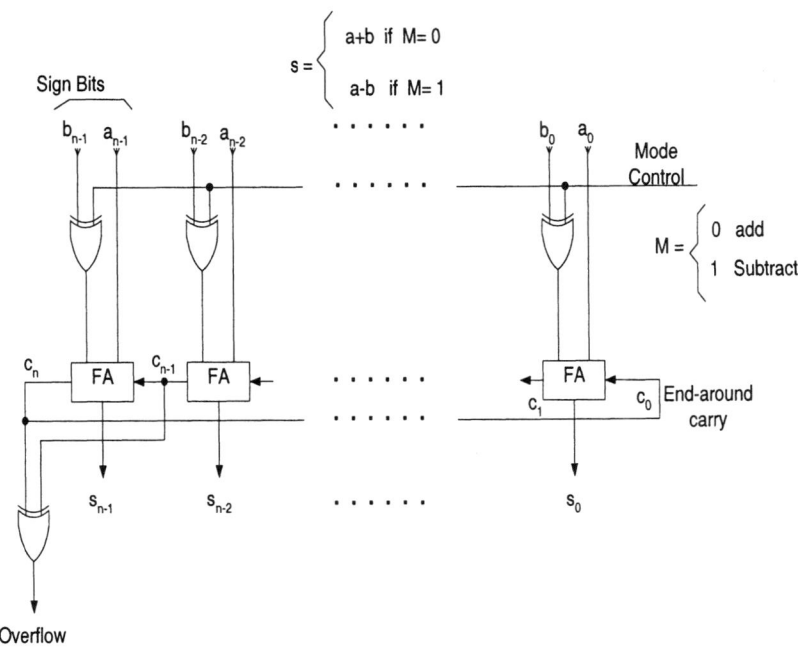

Fig. 2.11: One's Complement Addition/Subtraction

b) $|A| < |B|$.
Without the end-around carry the circuit performs

$$\begin{aligned} & 2^n - 1 \quad -|A| + |B| \\ = \ & 2^n - 1 \quad +(|B| - |A|) \\ > \ & 2^n - 1. \end{aligned}$$

That is, $c_n = 1$. After the *mod* 2^n function, the result generated by the circuit is $|B| - |A| - 1$ while the expected correct result is $|B| - |A|$. A "1" should be added to obtain the correct result, and $c_n = 1$ can provide such a "1".

So connecting the carry-out of MSB position to the carry-in of LSB position works correctly for all the cases. Whenever an add "1" operation is needed in the LSB position, the carry-out of MSB is equal to 1. Otherwise, it is equal to 0.

Figure 2.11 shows the schematic circuit of a 1's complement adder/subtractor. Comparing the area complexity of the 1's complement adder/subtractor with that of the 2's complement adder/subtractor, one can find that they are almost the same. As

48 ADDITION AND SUBTRACTION

far as the time complexity is concerned, we have the following calculation:

$$\Delta_T = \Delta_{XOR} + n\Delta_{FA}$$
$$+ n\Delta_{FA} \text{ (for end-around carry)}$$
$$= (2 + 2n + 2n)\Delta_g$$
$$= (4n + 2)\Delta_g.$$

In the 1's complement adder/subtractor, first c_n cannot be stable unless the carry rippled through all the FAs. Then c_n will be routed back to the LSB position and the second round of carry propagation will occur. So the delay time is doubled compared to the 2's complement adder/subtractor, and that's why the 1's complement adder/subtractor is not as popular as the 2's complement adder/subtractor.

2.5.3 Sign-Magnitude Addition

In the sign-magnitude addition, the two summands are given in the sign-magnitude form. After adding them together the sum should be in the sign-magnitude form as well. Figure 2.12 gives a block diagram of a sign-magnitude adder/subtractor. Detailed introduction will follow with attention paid to the pre-complement and post-complement schemes.

In the diagram a 1's complement adder is used to perform the addition. To provide the adder with operands in proper representation, a pre-complement operation may be required. Let's investigate the following cases exhaustively.

Augend	Op.	Addend		Augend	Op.	Addend	Pre-comp	$a_{n-1} \oplus M \oplus b_{n-1}$
$+\|A\|$	$+$	$+\|B\|$	$\Rightarrow$	$\|A\|$	$+$	$\|B\|$	No	0
$+\|A\|$	$+$	$-\|B\|$	$\Rightarrow$	$\|A\|$	$+$	$-\|B\|$	Yes	1
$+\|A\|$	$-$	$+\|B\|$	$\Rightarrow$	$\|A\|$	$+$	$-\|B\|$	Yes	1
$+\|A\|$	$-$	$-\|B\|$	$\Rightarrow$	$\|A\|$	$+$	$\|B\|$	No	0
$-\|A\|$	$+$	$+\|B\|$	$\Rightarrow$	$-(\|A\|$	$+$	$-\|B\|)$	Yes	1
$-\|A\|$	$+$	$-\|B\|$	$\Rightarrow$	$-(\|A\|$	$+$	$\|B\|\ )$	No	0
$-\|A\|$	$-$	$+\|B\|$	$\Rightarrow$	$-(\|A\|$	$+$	$\|B\|\ )$	No	0
$-\|A\|$	$-$	$-\|B\|$	$\Rightarrow$	$-(\|A\|$	$+$	$-\|B\|)$	Yes	1

Suppose A is positive. Since subtracting a number can be completed by adding the negated number, we have only two cases in the classification. Case 1: $|A| + |B|$, adding two summands of the same sign. Case 2: $|A|+(-|B|)$, adding two summands of different signs.

Suppose A is negative, the following two cases are considered. Case 1: $-(|A| + |B|)$, the magnitude of A and B should be added up, which is similar to Case 1 above. Case 2: $-(|A| + (-|B|))$, the magnitude of A and B cancel each other out, which is similar to Case 2 above.

When the magnitude of A and B cancel each other out we need to perform 1's complement operation on B's magnitude before the addition. This is called pre-complement. Let $P = a_{n-1} \oplus M \oplus b_{n-1}$, where a_{n-1} and b_{n-1} are the sign bits of

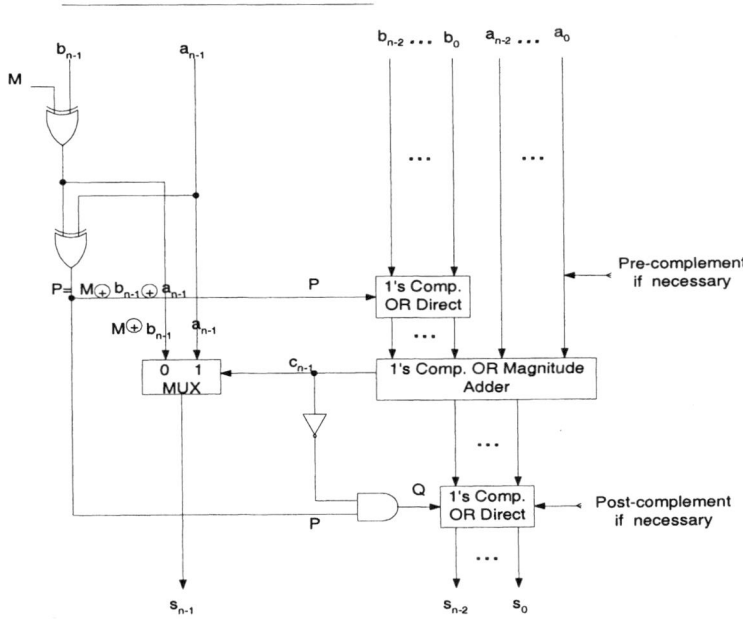

Fig. 2.12: Block Diagram of Sign-Magnitude Addition/Subtraction

A and B, respectively, and M is the mode in 0/1 for addition/subtraction. From the above analysis we can find that pre-complement is to be performed whenever $P = 1$. Hence, P is the enable signal for pre-complement.

Notice that when adding two numbers of different signs, if the magnitude of the negative number is greater, the result is negative. Also, adding two negative numbers results in a negative number. As the 1's complement adder represents its output in 1's complement format, we need to convert it to a sign-magnitude number. Therefore a post-complement is required.

The post-complement operation is enabled by $Q = P\bar{C}_{n-1}$. For case 1 examples listed in the above table, $P = 0$ and the post-complement is not needed. For case 2 examples, $P = 1$; the post-complement is needed only if $|B| \geq |A|$. Recall that the $(n-1)$-bit 1's complement adder performs $|A| + (-|B|)$ by $|A| + (2^{n-1} - 1 - |B|)$.

If $|B| \geq |A|$, subtracting a bigger number from and adding a smaller or equivalent number to $2^{n-1} - 1$ will result in a number $\leq 2^{n-1} - 1 < 2^{n-1}$. That is, $C_{n-1} = 0$. Hence, when $P = 1$ and $C_{n-1} = 0$, Q is asserted and a post-complement operation will be performed.

The sign of the result can be decided as follows. After converting the operation to addition if two summands have the same sign, the sign of the result can be equal to the sign of either of them. If the two summands have different signs, when $|B| \geq |A|$ ($c_{n-1} = 0$ case according to the above analysis) the sign of the result should be equal

50 ADDITION AND SUBTRACTION

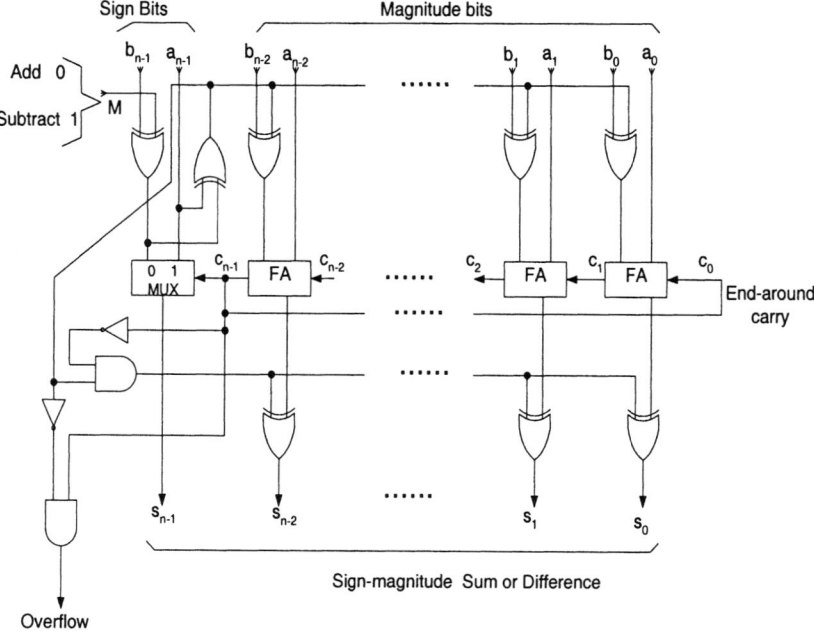

Fig. 2.13: Sign-Magnitude Addition/Subtraction

to the sign of addend. That is, $s_{n-1} = M \oplus b_{n-1}$. Otherwise ($c_{n-1} = 1$ case) $s_{n-1} = a_{n-1}$. By using C_{n-1} to control a 2-to-1 multiplexor which takes either a_{n-1} or $M \oplus b_{n-1}$, the sign of the result can be decided.

The schematic of an n-bit sign-magnitude adder is shown in Figure 2.13.

REFERENCES

1. I. Aleksander, "Array Networks for a Parallel Adder and Its Control," *IEEE Trans. Elec. Comp.*, Vol. EC-16. No. 2, Apr. 1967.

2. A. Avizienis, "Logic Nets for Carry and Borrow Propagation," Class Notes, Dept. of Engineering, University of California, Los Angeles, 1968.

3. B. E. Briley, "Some New Results on Average Worst-Case Carry," *IEEE Trans. Comp.*, Vol. C-22, No. 5, May 1973, pp. 459-463.

4. A. W. Burks, H. H. Goldstine and J. von Neuman, "Preliminary Discussion of the Logical Design of an Electronic Computing Instrument, Institute for Advanced

Study," Princeton, NJ, 1946 (reprinted in C. G. Bell and A. Newell, *Computer Structures: Readings and Examples*, McGraw-Hill, New York, 1971).

5. Fairchild Semiconductor Staff, *The TTL Applications Handbook*, Mountain View, CA, Aug. 1973.

6. D. Ferrari, "Fast Carry-Propagation Iterative Networks," *IEEE Trans. Comp.*, Vol. C-17, No. 2, Feb. 1968, pp. 132–145.

7. B. Gilchrist et al., "Fast Carry Logic for Digital Computers," *IRE Trans.* EC-4, Dec. 1955, pp. 133–136.

8. B. Gilchrist, J. H. Pomerene and S. Y. Wong, "Fast Carry Logic for Digital Computers, *IRE Trans. Elec. Comp.*, Vol. EC-4, no 4, Dec. 1955, pp. 133–136.

9. J. F. Kruy, "A Fast Conditional Sum Adder Using Carry Bypass Logic," *AFIPS Conf. Proceedings*, Vol. 27, FJCC 1965, pp. 695–703.

10. M. Lehman, and N. Burla, "Skip Techniques for High-Speed Carry-Propagation in Binary Arithmetic Units," *IRE Trans.* EC-10, No. 4, Dec. 1961, pp. 691–698.

11. M. Lehman, "A Comparative Study of Propagation Speed-Up Circuits in Binary Arithmetic Units," *Inform. Processing*, 1962, Elsevier-North Holland, Amsterdam, 1963, pp. 671–677.

12. H. Ling, "High-Speed Binary Parallel Adder," *IEEE Trans. Comp.*, EC-15, No. 5, Oct. 1966, pp. 799–802.

13. O. L. MacSorley, "High-Speed Arithmetic in Binary Computers," *Proc. IRE*, 49 (Jan. 1961), pp. 67–91.

14. G. W. Reitwiesner, "The Determination of Carry Propagation Length for Binary Addition," *IRE Trans.* EC-9, No. 1, Mar. 1960, pp. 35–38.

15. G. W. Reitwiesner, "Binary Arithmetic," in *Advances in Computers*, Vol. 1, F. L. Alt ed., Academic, New York, 1960, pp. 231–308.

16. S. Singh and R. Waxman, "Multiple Operand Addition and Multiplication," *IEEE Trans. Comp.*, Vol. C-22, No. 2, Feb. 1973, pp. 113–120.

17. J. J. Shedletsky, "Comment on the Sequential and Indeterminate Behavior of and End-Around-Carry Adder," *IEEE Trans. Comp.*, Vol. C-26, No. 3, Mar. 1977, pp. 271–272.

18. J. Sklansky and M. Lehman, "Ultimate-Speed Adders," *IRE Trans.* EC-12, No. 2, Apr. 1963, pp. 142–148.

19. P. M. Spira, "Computation Times of Arithmetic and Boolean Functions in (d,r) Circuits," *IEEE Trans. Comp.*, Vol. C-22, No. 6, June 1973, pp. 552–555.

20. C. W. Weller, "A High-Speed Carry Circuit for Binary Adders," *IEEE Trans. Comp.*, Vol. C-18, No. 8, Aug. 1969, pp. 728–732.

PROBLEMS

2.1 You are asked to implement $X = (A + \bar{B} + \bar{C}) \cdot (D + \bar{E}) \cdot \overline{FG}$. (a) Realize the logic based on the given expression strictly. (b) Improve your design applying the hard-wired AOI. (c) Find the delay time and area complexity for the designs in (a) and (b).

2.2 Construct a 10-to-1 multiplexer with three 4-to-1 multiplexers. The multiplexers should be interconnected and inputs labeled so that selection signals $s_3 s_2 s_1 s_0 = 0000$ through 1001 can be directly applied to the selection inputs without any added logic.
Hint: Make two levels of multiplexers. Partition $s_3 s_2 s_1 s_0$ so that some of them can make selection in level 1 and some in level 2. Note that data line 8 and 9 have only one bit different in the coding of their indexes.

2.3 Design the logic for a 3×4 ROM (with 3 address lines and 4 data lines), using a 3-to-8 decoder. Find the time complexity and area complexity for your design.

2.4 Given a 1's complement number $a_3 a_2 a_1 a_0$ including the sign bit, design a circuit to convert it to a sign-magnitude number. Find the area complexity of your circuit.

2.5 You are asked to design a 4-bit 2's complement number comparator. The two numbers to be compared are X and Y. The output signal GT (Greater Than) will be asserted if $X > Y$, LT (Less Than) will be asserted if $X < Y$ and EQ (Equal) will be asserted if $X = Y$. Show your logic expression for each output, and draw the schematic circuit. Note that the leading bit of each input operand is reserved for sign. Pay attention to the minimization of the area complexity.

2.6 Find the area complexity and delay time for the 2's complementer in Figure 2.10, assuming $n = 16$.

2.7 Find the delay time for a 32-bit 2's complement ripple carry adder. Show your work.

2.8 In a 1's complement adder, pay attention to the end-around carry. In which situation does the end-around carry = 1 and in which situation it = 0? Justify that the end-around carry will result in the right solution of addition.

2.9 Perform $-3 + 5$ applying a 4-bit sign-magnitude adder as shown in Figure 2.12. (a) What is the P value given by the circuit. Justify why it is the value you expect from the arithmetic point of view. (b) What is the output of the pre-complement? (c) What is the output of the full adder? (d) Repeat question (a) for Q. (e) Repeat question (b) for post-complement.

3
High-Speed Adder

3.1 CONDITIONAL-SUM ADDITION

In the study of ripple-carry adder we can see that carry propagation delay is the major concern when we try to speed up the addition of any two numbers. For an n-bit ripple-carry adder, the delay time is linear to n.

Notice that when adding two numbers as follows:

$$\begin{array}{r} a_i \\ b_i \\ +)\ \ c_i \\ \hline c_{i+1}\ \ \ s_i \end{array}$$

c_i can be either 0 or 1 in binary. Only two cases are possible:

$$\begin{array}{r} a_i \\ b_i \\ +)\ \ 0 \\ \hline c^0_{i+1}\ \ \ s^0_i \end{array} \quad \text{or} \quad \begin{array}{r} a_i \\ b_i \\ +)\ \ 1 \\ \hline c^1_{i+1}\ \ \ s^1_i \end{array}.$$

Here the superscript indicates the value of the carry-in assumed, and the subscript is for the bit index.

53

54 HIGH-SPEED ADDER

Under the assumption that the carry-in is 0, s_i^0 and c_{i+1}^0 can be easily found as follows:

$$s_i^0 = a_i \bar{b}_i + \bar{a}_i b_i = a_i \oplus b_i$$
$$c_{i+1}^0 = a_i b_i .$$

In a similar way, s_i^1 and c_{i+1}^1 can be found under the assumption that the carry-in is 1.

$$s_i^1 = a_i b_i + \bar{a}_i \bar{b}_i = a_i \odot b_i$$
$$c_{i+1}^1 = a_i + b_i ,$$

where $x \odot y$ is $\overline{x \oplus y}$.

Following the above logic, we build a conditional-sum adder. In a CS (conditional sum) cell, two sets of circuits can perform the additions based on two different carry-in values. Two sets of outputs can be prepared at the same time, and all that is left is to select one out of two once the actual carry-in value is known.

Using the carry-out of the lower order bit position as the carry-in to make selections in the higher order bit position, two bit positions are merged into one group. The selected sum of the higher order bit position concatenating that of the lower order bit position forms the sum of the group, and the carry-out of the higher order bit position becomes the carry-out of the group. In the next step two groups are to be merged together. The carry-out of the lower order group will be used to make selections in the higher order group.

In each step we merge every two adjacent groups at the same time, and the group size is doubled. First 1 bit is in a group, then 2, then 4 · · · · · · . In step k, 2^{k-1} bits are in a group. The carry-in to the LSB position or to the least significant group is always 0, which can rule out the other assumption and have the output of this group confirmed. As the size of the least significant group increases, more and more sum bits can be determined – through the selection in the higher order group to be merged rather than the carry propagation. The final result can be obtained in $\lceil log_2 n \rceil$ steps for n-bit long operands. In each step only two-to-one selections are needed which can be easily realized by multiplexers. Compared to the ripple-carry adder, which has a delay time linear to n, the conditional-sum adder achieves a significant improvement in time performance with an inexpensive hardware cost.

Given $A = (0110110)_2$ and $B = (1101101)_2$, Figure 3.1 shows the conditional-sum addition for illustration. A superscript 0 indicates that a carry 0 into each group is assumed, and a superscript 1 indicates that a carry 1 into each group is assumed. The groups are isolated from each other in the sense that the carry-out generated in a lower order group does not propagate to the higher order group. Between the pair of groups to be merged in the corresponding step, there are arrow(s) pointing from the right group to the left group. It points away from the carry-out of the right group which is used to make a selection in the left group. If that carry is 0 it points to the upper set of data in the left group. If that carry is 1 it points to the lower set of data. Only in the least significant group such carry-out has a known value (as in the circle), and the sum it picks up contributes to the final result (as in the box). Otherwise

Step	i	6	5	4	3	2	1	0
	a_i	0	1	1	0	1	1	0
	b_i	1	1	0	1	1	0	1
1	S^0_i	1	0	1	1	0	1	1
	C^0_{i+1}	0	1	0	0	1	0	0
	S^1_i	0	1	0	0	1	0	
	C^1_{i+1}	1	1	1	1	1	1	
2	S^0_i	0	0	1	1	0	1	1
	C^0_{i+1}	1		0		1		0
	S^1_i	0	1	0	0	1	0	
	C^1_{i+1}	1		0		1		
3	S^0_i	0	0	1	1	0	1	1
	C^0_{i+1}	1				1		
	S^1_i	0	1	0	0			
	C^1_{i+1}	1						
Result	S	0	1	0	0	0	1	1
	C_{out}	1						

Fig. 3.1: Conditional-Sum Addition

we see two arrows from each group on the right representing two carry-out signals participating in the selection operation. They all have assumed values at this moment and none knows what the actual carry is. The candidates they select in the left group may or may not constitute the final result. The route of the signals are provided in the circuit, however, and a unique set of data will be selected when merged into the least significant group in a later step. Some right most groups in the figure are incomplete groups in terms of their size. Only the "left over" bit positions are contained for the case that n is not a power of 2.

The schematic circuit of a conditional-sum adder is given in Figure 3.2 to add the above 7-bit numbers.

In (b), the data routed through the adder are shown given the particular inputs.

3.2 CARRY-COMPLETION SENSING ADDITION

The linear delay time of the ripple-carry adders is due to the carry propagation from the least significant bit position to the most significant bit position. That is the time we need to wait in the worst case for the FA in MSB position to produce the correct sum and carry-out. Notice that composed of combinatorial circuit, FA will produce something as long as data present on its input lines. The input signal may change later, and so does the output in accordance. Before the worst case delay time is reached, the correctness of the output cannot be guaranteed and we cannot claim the completion of the addition.

In this section we introduce a carry-completion sensing adder (CCSA). It allows multiple carries to be propagated independently and simultaneously, and senses the completion of all the propagation. On "average" it takes shorter time to make the output stable, and such adder is considered <u>faster</u> than the conventional ripple-carry adder.

Two types of carries, independent carry (IC) and dependent carry (DC), are defined as follows. Note that each type of carries can be initiated at all the bit positions simultaneously without waiting for the propagated carries from the preceding bit positions.

$$\left. \begin{array}{l} a_i = 0 \\ b_i = 0 \end{array} \right\} IC_{i+1} = 0$$

$$\left. \begin{array}{l} a_i = 1 \\ b_i = 1 \end{array} \right\} IC_{i+1} = 1$$

$$\left. \begin{array}{l} a_i = 0 \\ b_i = 1 \\ or \\ a_i = 1 \\ b_i = 0 \end{array} \right\} DC_{i+1} = \begin{cases} DC_i & \text{if it exists} \\ IC_i & \text{otherwise} \end{cases}$$

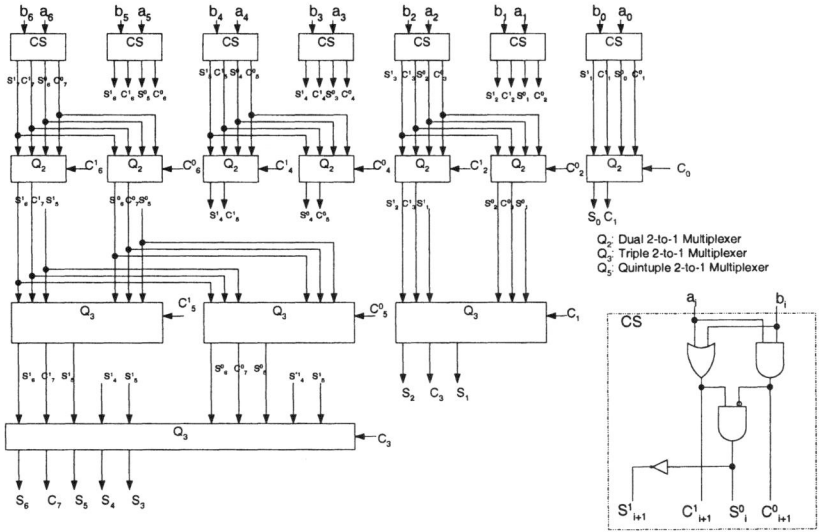

(a) Circuit of Conditional-Sum

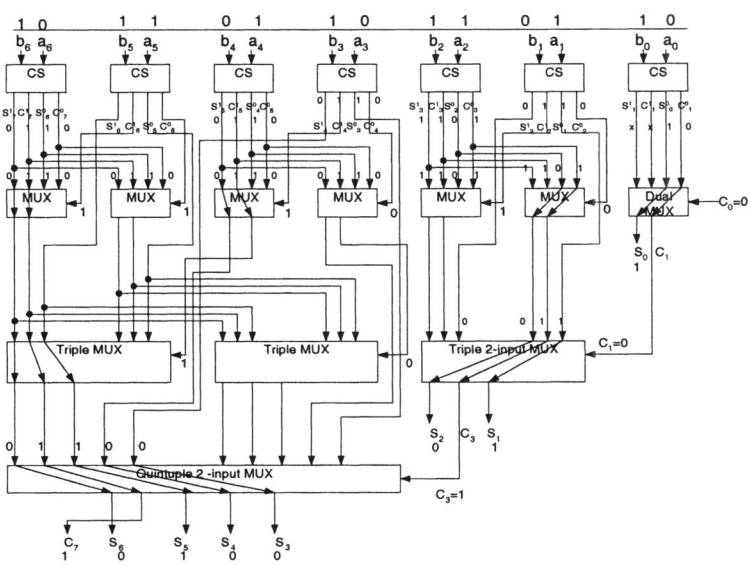

(b) Data Routed through the Adder

Fig. 3.2: Conditional-Sum Adder

58   HIGH-SPEED ADDER

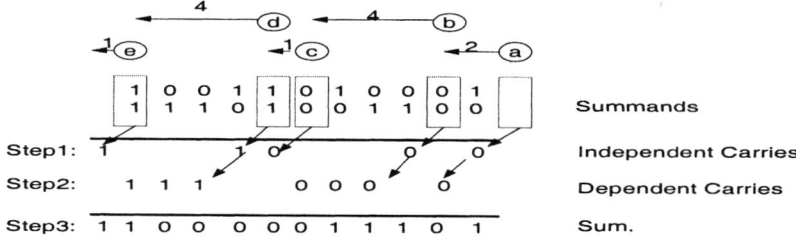

Fig. 3.3: Generation and Transmission of Carries

Here IC_{i+1} can be determined based on the two summand bits a_i and b_i. Since all the a_is and b_is are available at the same time, all the independent carries can be generated in parallel.

On the other hand, DC_{i+1} cannot be determined unless DC_i or IC_i is known, and DC_i is based on DC_{i-1} or IC_{i-1}. Here, starting from the independent carries, multiple carry propagation can go on. Ending at the next higher bit position in which the independent carry exists, some carry propagation is longer while some is shorter. Refer to the numerical example in Figure 3.3 showing the independent/dependent carries and the carry propagation. The blocked bit position can generate the carry, and others can propagate the carry triggled by either the independent or the dependent carry of the lower order bit position.

In Step 1, the existence of independent carries for any particular column is determined by detecting the equality of the pair of summand bits in the preceding column; these independent carries are then set to the value as defined above. In Step 2, the dependent carries are generated by setting all the carries between each pair of independent carries equal to the right member of the pair. In Step 3, the carry and the summand bits in each column are added with module-2. That is, in the bit-wise addition the carry-out resulted from any bit position is neglected and not taken into consideration in the higher order bit position.

Define the *carry propagation length* as the length of the propagation since the carry was triggled until it stops propagating. We can see in this example the various lengths of carry propagation among which the longest is 4. It has been proved statistically that the upper bounded of the average longest carry propagation length in random binary numbers is $\lceil log_2 n \rceil$.

Figure 3.4 shows the logic circuit of a carry-completion sensing adder. A *carry transmission* (CT) cell is given in (a). If $a_i b_i = 11$, the output of gate 1 will be asserted and the carry-out $c_{i+1} = 1$. If $a_i b_i = 00$, the output of gate 2 will be asserted and the carry-out $\bar{c}_{i+1} = 1$. These two cases are exclusive, and gate 1 and gate 2 can have only one outputting 1. That means when $c_{i+1} = 1$ there must be $\bar{c}_{i+1} = 0$, and when $\bar{c}_{i+1} = 1$ there must be $c_{i+1} = 0$. The carry generated in these two cases is the independent carry.

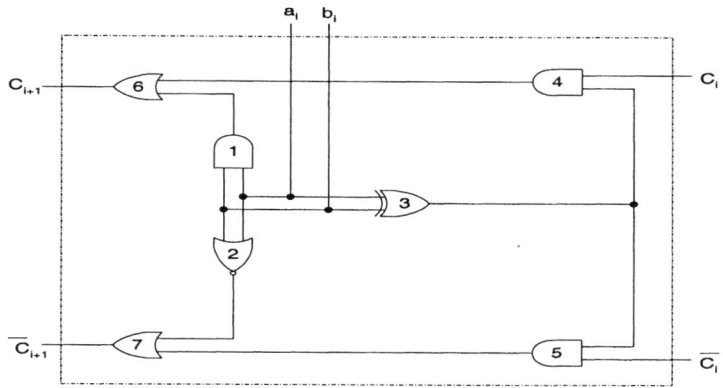

(a) Carry Transmission Unit

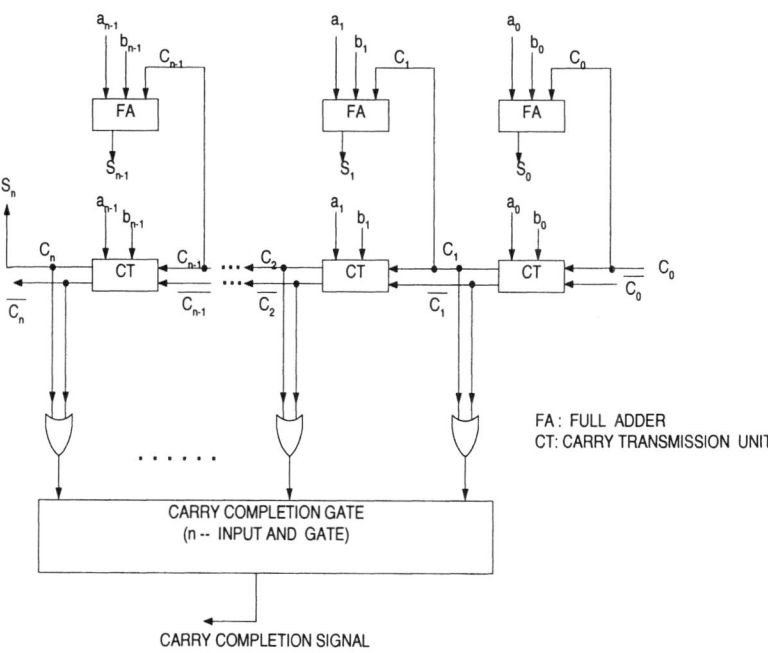

(b) Carry-Completion Sensing Adder

Fig. 3.4: Construction of Carry-Completion Sensing Adder

If $a_i b_i = 01$ or 10, the output of gate 3 is asserted, and that of gate 1 or gate 2 is not. Then, depending on the carry-in c_i and $\bar{c}_i$, either gate 4 outputs a 1 and gate 5 outputs a 0 or vice versa. Such signals will be transmitted through gate 6 and 7 to the outputs c_{i+1} and $\bar{c}_{i+1}$, becoming the carry-out signals – dependent carry. Note that this case and the previous two cases are mutually exclusive, that is, only one of them can occur at each bit position.

For a bit position in which the two summand bits $a_i = b_i$, a new carry, independent carry 1 or 0, starts propagation at once. In this same bit position, the existing carry propagation from the lower order bit position stops. For the bit position in which $a_i \neq b_i$, neither gate 6 nor gate 7 can output a 1 before the arrival of c_i and $\bar{c}_i$. One of the outputs, c_{i+1} or $\bar{c}_{i+1}$, should be high to show the carry-transmission completed in bit position i. c_{i+1} and $\bar{c}_{i+1}$ are ORed together in each bit position, and then sent to an n-input AND gate. The AND gate is a carry-completion sensing gate whose output is high if and only if all the inputs are 1, indicating that all the bit positions have the carry generation and propagation completed.

The logic of the CT cell depicted in the figure can be expressed as follows:

$$c_{i+1} = a_i \cdot b_i + (a_i \oplus b_i) \cdot c_i$$
$$\bar{c}_{i+1} = \bar{a}_i \cdot \bar{b}_i + (a_i \oplus b_i) \cdot \bar{c}_i$$

and

$$c_0 = 0$$
$$\bar{c}_0 = 1,$$

where c_0 is the initial independent carry into the LSB position. Note that c_{i+1} in the above equation is composed of the generation of the independent carry, and the generation and propagation of the dependent carry.

The time complexity of this adder is composed of three major components, but they may be overlapping each other:

1. Time for all independent carries generation.

2. Time for dependent carries generation <u>and</u> propagation.

3. Time for the final piecewise summation which gives the final answer.

Refer to the numerical example in Figure 3.3 with $n = 11$, and the circuit in Figure 3.4. In the CT of all the bit positions after the summand bits a_i and b_i are input, gates 1, 2 and 3 can have the outputs ready within no longer than $2\triangle_g$ time. Then from c_i to c_{i+1}, each bit position needs $4\triangle_g$ to propagate the carry, and $\lceil log_2 n \rceil$ bit positions should be considered according to the average propagation length. Then the delay of OR gate and the delay of the n-input AND gate should be added with which the $\triangle_{FA}$ is overlapped. Here the circuit is an asynchronous one. The FA always outputs something with some value present on c_i, the carry-in, though it may not be a correct output. Until the AND gate transmits to the control circuitry a signal indicating that all the carry generation and propagation have been completed, the outputs of the FAs cannot be adopted for further utilization.

Using the upper bounded value we have

$$\begin{aligned}
\triangle_{CCSA} &= \triangle_{CT} + \triangle_{OR} + \triangle_{AND} \\
&= (2\triangle_g + (\lceil log_2 n \rceil) \cdot 4\triangle_g) + 2\triangle_g + 2\triangle_g \\
&= (6 + 4\lceil log_2 n \rceil)\triangle_g .
\end{aligned}$$

$$\begin{aligned}
A_{CCSA} &= n(A_{FA} + A_{CT} + A_{OR}) + A_{AND} \\
&= n[A_{FA} + (A_{XOR} + 3A_{AND} + 2A_{OR} + A_{NOR}) + A_{OR}] \\
&\quad + A_{AND} \\
&= \{n[10 + (3 + 6 + 4 + 1) + 2] + 2\}A_g \\
&= (2 + 26n)A_g.
\end{aligned}$$

3.3 CARRY-LOOKAHEAD ADDITION (CLA)

In the ripple-carry adder, the carries in different bit positions are generated sequentially. That is, C_{i+1} is dependent on C_i, and C_i cannot be determined unless C_{i-1} is known. We introduce in this section an adder which can generate all the carries in parallel. No carry propagation is in the cause of delay.

3.3.1 Carry-Lookahead Adder

Let $A = (A_{n-1} \cdots A_1 A_0)$ and $B = (B_{n-1} \cdots B_1 B_0)$ be the augend and addend, respectively, and C_i the carry into bit position i. The carry into LSB is C_0. Investigating the condition that a carry can be generated in bit position i neglecting the carry-in C_i, we define the *carry generation* function

$$G_i = A_i \cdot B_i.$$

The condition that position i can pass the carry-in from position $i-1$ to position $i+1$ can be defined as *carry propagation* function

$$P_i = A_i \oplus B_i.$$

Hence S_i and C_{i+1} can be expressed as the functions of P_i and G_i.

$$\begin{aligned}
S_i &= (A \oplus B) \oplus C_i \\
&= P_i \oplus C_i
\end{aligned}$$

and

$$\begin{aligned}
C_{i+1} &= A_i \cdot B_i + (A \oplus B) \cdot C_i \\
&= G_i + P_i \cdot C_i.
\end{aligned}$$

62 HIGH-SPEED ADDER

Again, our goal here is to speed up the carry propagation.

P_i and G_i, for $i = n-1, \cdots, 1, 0$, can be generated in parallel upon A_is and B_is which are available at the very beginning of the addition, (see part (a) of Figure 3.5). By applying the above equation recursively, all the C_{i+1}s can be generated based on G_i, P_i and C_0.

$$\begin{aligned}
C_1 &= G_0 + C_0 P_0 \\
C_2 &= G_1 + C_1 P_1 \\
    &= G_1 + G_0 P_1 + C_0 P_0 P_1 \\
C_3 &= G_2 + C_2 P_2 \\
    &= G_2 + G_1 P_2 + G_0 P_1 P_2 + C_0 P_0 P_1 P_2 \cdots \\
C_{i+1} &= G_i + G_{i-1} P_i + G_{i-2} P_{i-1} P_i + \cdots + G_0 P_1 P_2 \cdots P_i \\
        &\quad + C_0 P_0 P_1 \cdots P_i \cdots \\
C_n &= G_{n-1} + G_{n-2} P_{n-1} + \cdots \\
    &\quad + G_0 P_1 P_2 \cdots P_{n-1} + C_0 P_0 P_1 \cdots P_{n-1}. \quad (3.1)
\end{aligned}$$

Notice that C_{i+1} is independent of C_i now.

The circuit to generate C_{i+1}s is called carry-lookahead unit shown in part (b) of Figure 3.5.

The final sum S can be computed once C_is, $i = 1, 2, \cdots, n$, are known. See part (c) in Figure 3.5 for the circuit.

The delay time of the carry lookahead adder can be calculated as follows:

$$\begin{aligned}
\triangle_T &= \triangle_{carry\ gener./prop.\ unit} + \triangle_{CLA} + \triangle_{SUM} \\
&= 3\triangle_g + 2\triangle_g + \triangle_{XOR} = 7\triangle_g.
\end{aligned}$$

One can see that in Equation (3.1), the number of terms in the OR function is as big as $n+1$, and in the last term the number of valuables in the AND function is also $n+1$. As the number of bits n increases, the fan-in and fan-out could be a problem if carry-lookahead is adopted. On the other hand, sequential generation of each C_{i+1} is too slow. As a combination of the above two approaches, block carry lookahead adder (BCLA) can be adopted in which group(s) of carries are generated in parallel.

3.3.2 Block Carry Lookahead Adder

Same as before, in a block carry lookahead Adder (BCLA) the carry generation and propagation variables, P_is and G_is are prepared simultaneously for $i = 0, 1, \cdots, n-1$.

Partition the n bits into $\frac{n}{m}$ blocks with m bits in each block. We define the *block carry generate* function G^* and *block carry propagate* function P^* for an m-bit block as follows

$$\begin{aligned}
P^* &= P_0 P_1 \cdots P_{m-1} \\
G^* &= G_{m-1} + G_{m-2} P_{m-1} + G_{m-3} P_{m-2} P_{m-1} + \cdots \\
    &\quad + G_1 P_2 P_3 \cdots P_{m-1} + G_0 P_1 P_2 P_3 \cdots P_{m-1}.
\end{aligned}$$

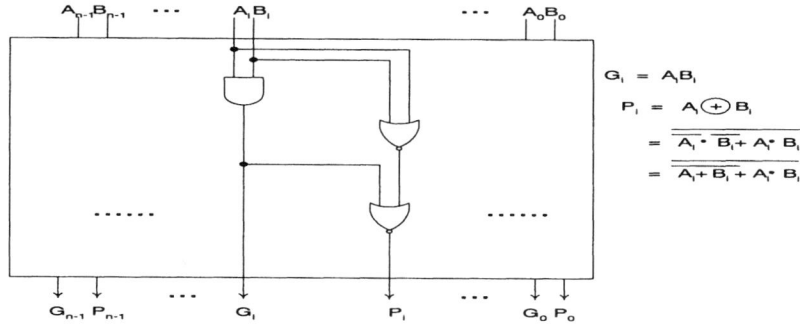

(a) Carry Generate/Propagate Unit

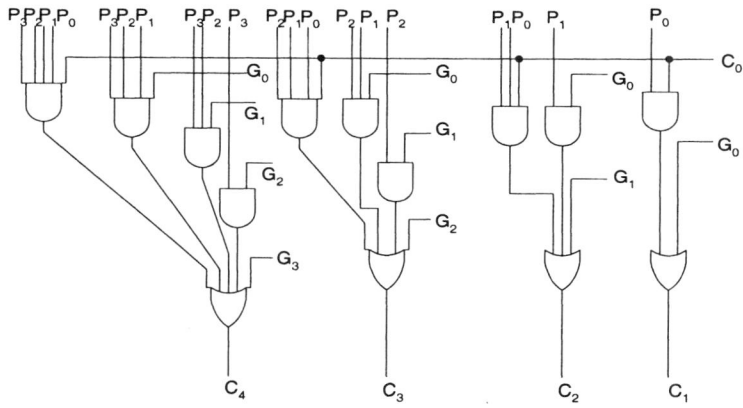

(b) Carry-Lookahead Unit

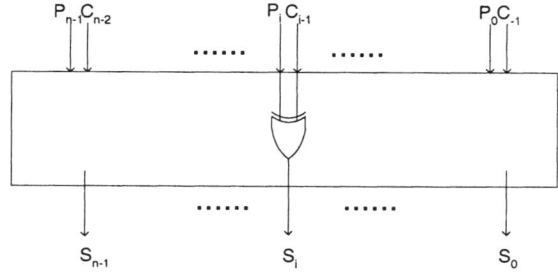

(c) Summation Unit

Fig. 3.5: Carry-Lookahead Adder

$P^* = 1$ if the block can propagate a carry. It is not difficult to understand that if every bit position in the block can propagate a carry, then the block is capable of carry propagation. $G^* = 1$ if the block can generate a carry. One can see that if some bit position in the block can generate a carry and every bit position of higher order than it can propagate the carry, then this block is capable of carry generation.

The schematic circuit of a 4-bit BCLA unit is shown in Figure 3.6(a). We can see that the P^* and G^* of each block can be generated as long as P_is and G_is are available. On the other hand, the carry-out C_{i+1} for bit position i in the block is dependent not only on P_is and G_is, but also on the carry-in to the block (C_0 for the first block and C_{4k} for block k in general). We discuss next how to provide such carry-in to each block.

Here another level of carry lookahead is needed. Recall that in Equation (3.1), given G_is and P_is, we can find carry-out C_{i+1}s for bit position i in parallel. Similarly, given block carry propagation P^*s and block carry generation G^*s, we can find in parallel the carry-out for each block such as $C_4, C_8, \cdots\cdots, C_{16}$, with $m = 4$. In general $C_{m(k+1)}$, $k = 0, 1, \cdots, \frac{n}{m} - 1$, is the carry-out of block k which can be used as the carry-in of block $(k + 1)$. After we feed it back to BCLA $(k + 1)$ of the first level carry lookahead, $C_{4(k+1)+1}$, $C_{(4k+1)+2}$ and $C_{4(k+1)+3}$ can be generated in parallel.

Here there are two levels of carry lookahead. All the m carries out of the same block are generated in parallel, and all the carry-in signals to $\frac{n}{m}$ blocks are generated in parallel. The $\frac{n}{m}$ carries are generated first, then the m carries follow each of them. These two types of carries are generated one type after another in sequence.

Figure 3.6(b) shows a block carry lookahead adder performing a 16-bit addition. The boxes in the two levels of carry-lookahead, BCLA and CLA Unit, can be replaced by the circuit in (a), where P_is and G_is are input to the upper level, and P^*s and G^*s are input to the lower level. P^*s and G^*s are the output of the upper level carry-lookahead only. The unit at the top is the same as in Figure 3.5(a). The unit at the bottom generates the summation bits in parallel once all the C_{i+1}s are available. Recall that

$$S_i = P_i \oplus C_i$$
$$= (A_i \oplus B_i) \oplus C_i.$$

The delay time of the two-level BCLA can be found by

$$\triangle_T = \triangle_{carry\ gener./prop.\ unit} + \triangle_{BCLA^1} + \triangle_{4-bit\ CLAU} + \triangle_{BCLA^2} + \triangle_{SUM}$$
$$= 3\triangle_g + 2\triangle_g + 2\triangle_g + 2\triangle_g + 2\triangle_g = 11\triangle_g$$

if the AND-OR function is converted to a NAND-NAND realization. Notice that two passes of BCLA delay have been considered. In each pass a different data route is taken, though the amount of time spent on the routing may remain the same. The first pass $BCLA^1$ is from P_is/G_is to P^*s/G^*s, and the second pass $BCLA^2$ is from C_{4k} to C_{4k+3}. Such an idea can be extended to involve more than two levels of BCLA into an adder. Each additional level of carry-lookahead will contribute a $4\triangle_g$ extra delay time, $2\triangle_g$ for $BCLA^1$ and $2\triangle_g$ for $BCLA^2$.

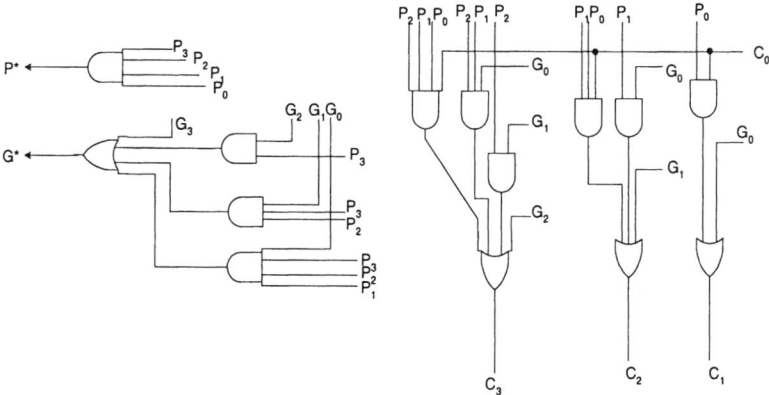

(a) BCLA Circuit

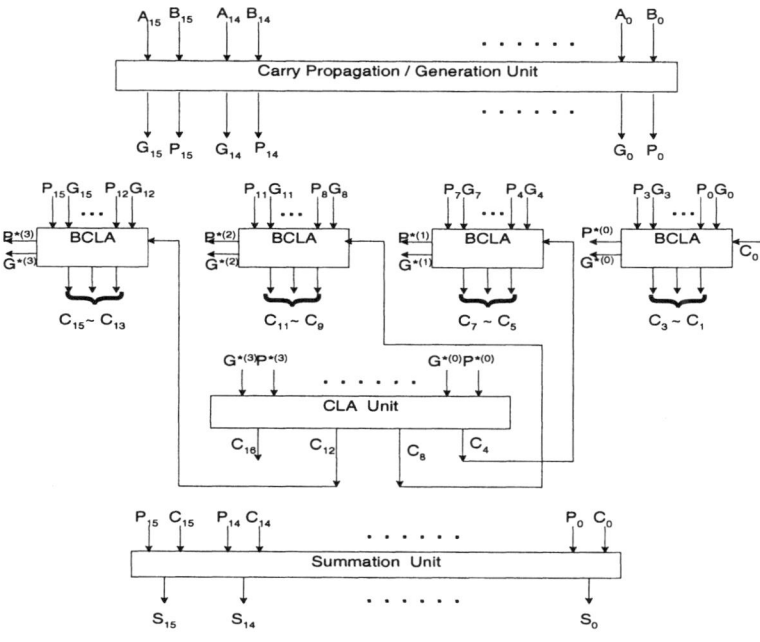

(b) Two-Level Carry-lookahead

Fig. 3.6: Block Carry-Lookahead Adder

3.4 CARRY-SAVE ADDERS (CSA)

The various types of adders we have discussed so far can add two numbers only. In array processing and in multiplication and division, multioperand addition is often encountered. More powerful adders are required which can add many numbers instead of two together. We introduce in this section the design of a high-speed multioperand adder called a carry-save adder (CSA). Instead of waiting for the carry propagation of the first addition to be complete before starting the second addition, the idea here is to overlap the carry propagation of the first addition with the computation in the second addition, and so forth, since repetitive additions will be performed by a multioperand adder. After the last addition, the carry propagation delay is then unavoidable and it should be included in the total delay time.

Figure 3.7 is a schematic diagram of an n-bit CSA composed of n full adders. Registers are used to temporarily buffer the outputs of the full adders before sending them to the input lines of the adders. This kind of register is called the *carry-save register*.

Let F^1, F^2, $\cdots$, F^k be the k summands to be added up, and X, Y, Z be the three inputs of the carry-save adder. First, load numbers F^1, F^2 and F^3 into the adder and add. Second, the results S and C are fed back to X and Y, and F^4 is loaded into Z. F^4 is added with the saved carry and the partial sum obtained from the previous addition. The second operation will be repeated with a new summand taken each time until F^k is taken.

Let

$$F^j = (f^j_{n-1} f^j_{n-2} \cdots f^j_0),\ 1 \leq j \leq k,$$

be the k numbers to be added. The operations for each full adder in bit position i are as follows.

1. Input f^1_i, f^2_i and f^3_i to the full adder and add. Store the result s_i in R_i and c_{i+1} in R'_i.

2. Feed the register-stored s_i back to the full adder in the current bit position as its first input; send the register stored c_{i+1} to the full adder in bit position $(i+1)$. The full adder in the current bit position takes the carry-in from bit position $i-1$ as its second input and takes f^4_i from a new summand as its third input and adds. Store the result s_i in R_i, and c_{i+1} in R'_i.

3. Repeat Step 2, but rather than f^4_i, take f^{j+3}_i in the jth iteration as the third input of the full adder and add until $j = k$.

4. Repeat Step 3, but take 0 as the third input of the full adder until all the registers $R'_{n-1}, \cdots, R'_0$ contain 0s.

In step 1 to Step 3, repeat $(k-2)$ additions for adding k numbers since three numbers can be added up in the first cycle and the rest $(k-3)$ numbers need $(k-3)$ cycles to add them up.

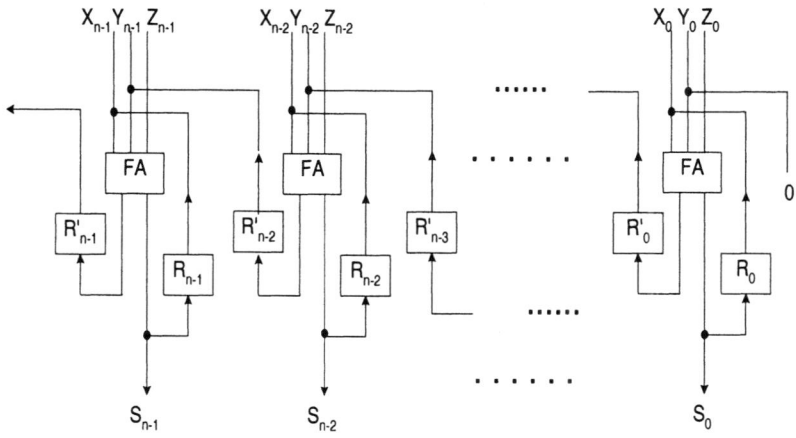

(a) *n*-bit Carry-Save Adder

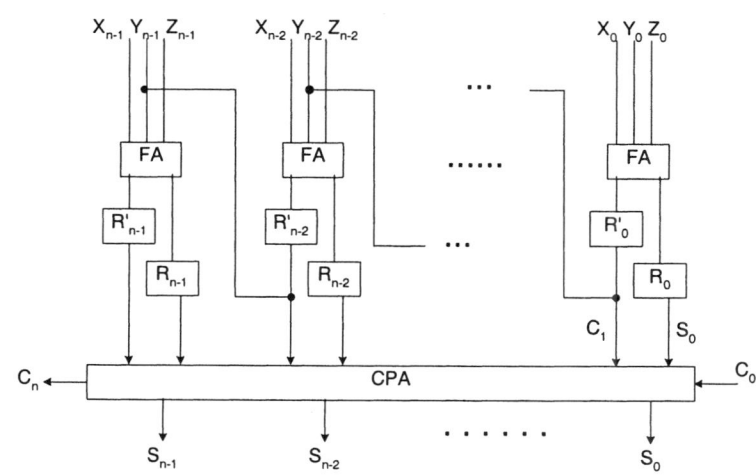

(b) Applying CPA to Add Carry Vector and Sum Vector

Fig. 3.7: Carry-Save Adder

68 HIGH-SPEED ADDER

After Step 3, the carry propagation takes place. Within no more than $n - 1$ cycles, in addition to the $(k - 2)$ cycles mentioned above, the final $Sum = (S_{n-1} \cdots S_0)$ can be available.

Let the total time steps for adding k n-bit numbers be t cycles. We have

$$k - 2 \leq t \leq n + k - 3.$$

The delay time of the CSA is

$$\triangle_{CSA} = t \times f(\triangle_{FA}, \triangle_{FF}).$$

The area complexity of an n-bit CSA is

$$A_{CSA} = n \times f(A_{FA}, A_{FF}).$$

When the word length n is very long, the ripple carry propagation in the final stage will significantly degrade the performance of CSA. To speed up Step 4, a carry propagate adder (CPA) can be used. (See the block diagram in Figure 3.7(b).) Since in Step 4 only two vectors, the sum vector and the carry vector, are to be added, the CPA can be any kind of high-speed two-operand adder described in prior sections of this chapter. For example, a CLA can be adopted. In that case the required number of cycles to complete the addition of k numbers will be

$$t = k - 2 + 1 = (k - 1) \; cycles.$$

The previously described CSA is of single level. Recall that only one number can be added in Step 2. That is, $N^4, N^5, \cdots, N^k$ are added sequentially. We can connect multiple levels of CSAs in a tree fashion to add k numbers simultaneously where $k = 4, 5, 6$ and 7.

Given in Figure 3.8(a) is a 5-input CSA tree. Notice that each input or output line represents n single-bit lines. Each CSA block includes n full adders with no flip-flops. The left pointed arrows on the carry output lines indicate that the carries are shifted left for one bit position before being fed to the next stage and a 0 is entered into the LSB position. At the bottom of the tree, a CPA (not shown in the figure) is required to add the sum vector and carry vector together. If the number of inputs is a multiple of 3, an alternative tree structure can present. Figure 3.8(b) shows an example of a 6-input CSA tree.

To add k numbers, $(k - 2)$ CSAs are required with each additional input operand increasing the number of CSAs by one. Hence the area complexity of the CSA tree is

$$A_{CSA \; Tree} = (k - 2)A_{CSA} + A_{CPA}.$$

The total delay time of a CSA tree is dependent on the number of levels in the tree. Each level of CSAs contributes $2\triangle_g$ to the propagation delay which is the delay time of a full adder. Hence

$$\triangle_{CSA \; Tree} = 2l\triangle_g + \triangle_{CPA},$$

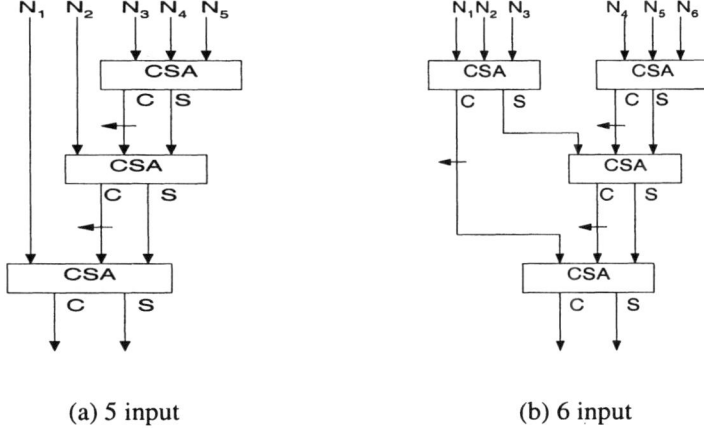

(a) 5 input (b) 6 input

Fig. 3.8: Carry-Save Adder Tree

where l is the number of levels.

Let $\lambda(l)$ be the maximal number of operands which can be added by an l-level CSA tree. $\lambda(1) = 3$. Each CSA has 3 inputs and 2 outputs, hence the number of outputs times $\frac{3}{2}$ will be the number of inputs, that is, the number of outputs in the upper level. If the number of outputs is not a multiple of 2, then it mod 2 indicates the number of extra outputs in the upper level. Hence $\lambda(l)$ can be defined recursively as follows:

$$\lambda(l) = \left\lfloor \frac{\lambda(l-1)}{2} \right\rfloor \times 3 + \lambda(l-1) \ mod \ 2.$$

For different values of l, the maximal numbers of input operands that a CSA tree can add are listed in Table 3.1.

There are two ways to conduct the multi-operand addition. In one method, all the bits of an operand are processed in parallel, and the operands are taken one after another. In the other method, all the bits in the same position of operands are admitted together. The examination goes over one bit position after another. See Figure 3.9 for reference. The "×" represents an individual bit. The carry-save addition previously introduced belongs to the first method. What is introduced in the following section is an addition using the second method, bit partitioned addition. Here the carry-save adders perform the multioperand addition in a row-wise fashion, while the *bit partitioned adders* to be introduced in this section add k numbers in a column-wise approach.

70 HIGH-SPEED ADDER

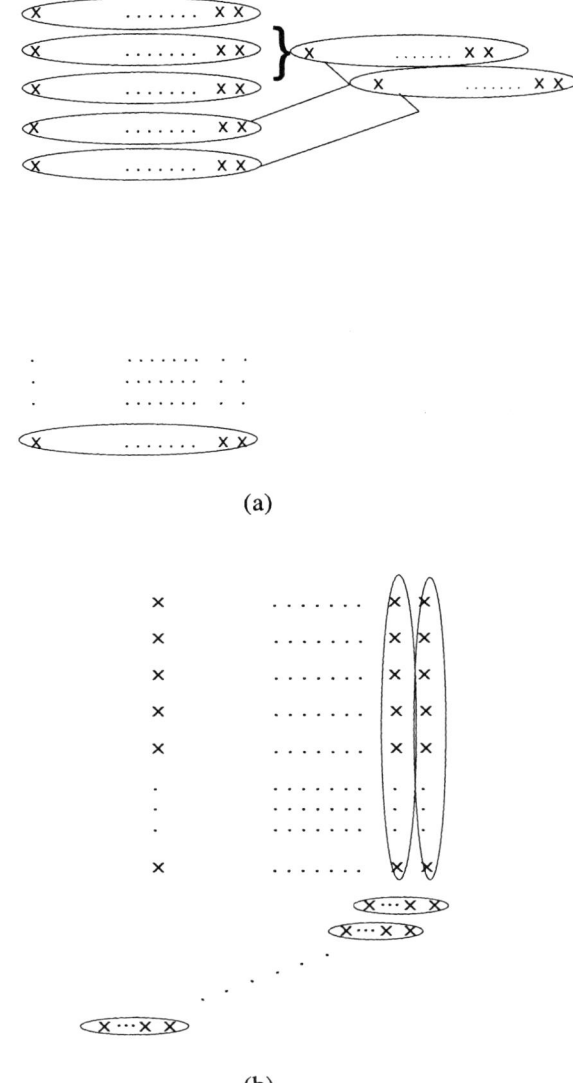

Fig. 3.9: Two Types of Parallelization in Multi-Operand Addition

Table 3.1: Maximum Inputs of CSA Trees

l	$\lambda(l)$
1	3
2	4
3	6
4	9
5	13
6	19
7	28
8	42
9	63

3.5 BIT-PARTITIONED MULTIPLE ADDITION

In the bit partitioned multiple addition, the numbers are added column by column with all the adders taking care of one bit at a time. The count of "1"s in a particular column is represented in binary numbers that are added up, with each displaced one bit position from its neighbor. Depending on the number of multiple operands to be added, say k, the partial sum length is $\lceil log k \rceil$.

The design of bit-partitioned adders is illustrated with an example of adding seven 4-bit numbers shown in Figure 3.10. The seven numbers are stored in a register and are shifted left one bit position at a time. In this way the seven numbers are fed into a 7-input *column adder* (CA) column by column with the leftmost column first.

Observing that the maximal sum of a 7-bit addition is 7 which can be represented by 3 bits, we reserve three output lines for the CA. The column adder can be implemented with a ROM or PLA – a device which can produce a 3-bit data given a corresponding 7-bit pattern. (There are in total 2^7 of such patterns.)

The sums of different bit-slices will be added up sequentially. A "carry-save" like adder is used for this purpose. Note that different weights should be given to the CA results generated for different bit-slices. The partial sum obtained after the addition of the bit i slice should be shifted left for one bit position before the bit $(i - 1)$ slice can be added on. Hence one can see that instead of feeding the old sum bit to the same bit position as in a typical carry save adder, it is saved now and later fed to the left bit position in the next cycle. Since some bit positions have only two input bits to add up, half-adders can be used to replace some full adders.

The delay time of the bit-partitioned adder is proportional to the word length of the operands. To add numbers of n bits, n cycles are needed which is apparently independent of k. However, with a closer look we can find that the period of each cycle is related to k. For example, the ROM size in CA increases as k increases, so does the delay along the chain of HAs and FAs.

72 HIGH-SPEED ADDER

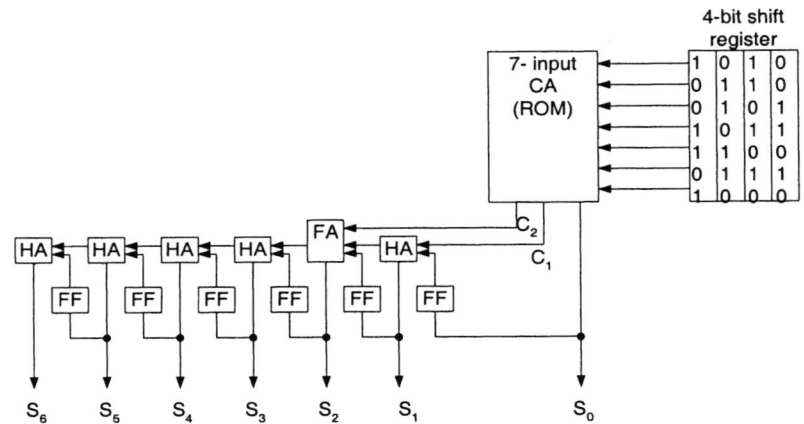

(a) Bit-Partitioned Adder

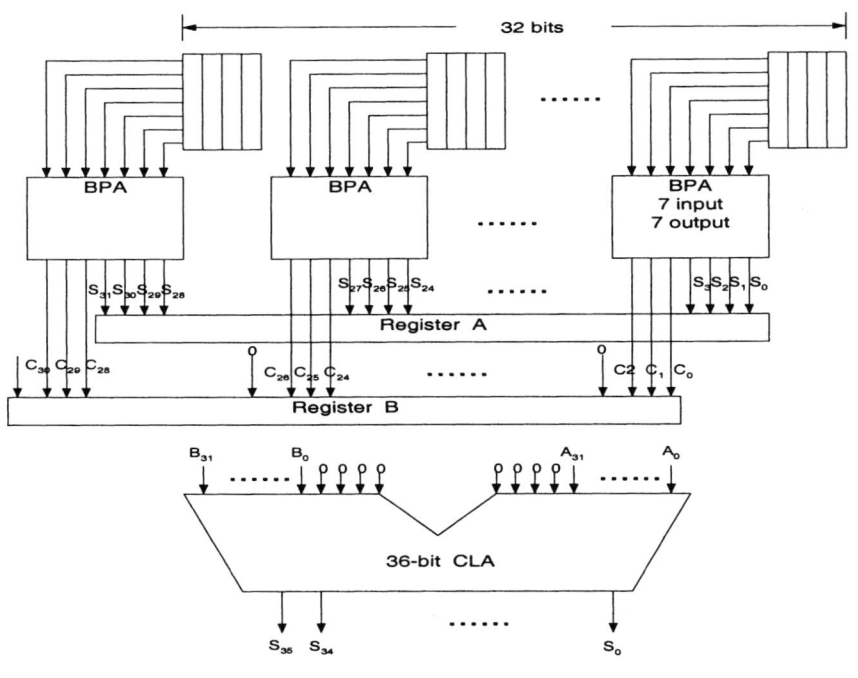

(b) Block Bit-Partitioned Adder

Fig. 3.10: Bit-Partitioned Multiple Addition

When n is large, the adder can be further partitioned into "groups" with each group containing m bit slices. In Figure 3.10(b), to perform a full length (32 bits) seven-number addition, eight of the 4-bit partitioned subadders (PSA) are used, each generating a 7-bit partial result. Since the adjacent groups have a bit position offset of 4, every 4 bits from the corresponding positions of each group can be concatenated to form a full length number. We feed the lower order 4 bits generated by each group to register A and the higher order 3 bits from each group to register B, and then perform a two-operand addition $A + B$. A fast two-operand adder, CLA, is adopted to complete the addition. Notice that since the LSB (and MSB, respectively) of number B is 4 bit positions higher than the LSB (MSB) of number A, four 0s are entered to the least significant bit positions of the addend and four are entered to the most significant bit positions of the augend for lineup.

The delay time of the above adder can be expressed as

$$\triangle_T = m \; cycles + \triangle_{CLA} = (m + 1) \; cycles$$

in general. When $m=3$, $\triangle_T = 4 \; cycles$. Again, the total delay time is independent of the number of input operands k.

REFERENCES

1. D. P. Agrawal and T. R. N. Rao, "On Multiple Operand Addition of Signed Binary Numbers," *IEEE Trans. Comp.*, Vol. C-27, no. 11, Nov. 1978, pp. 1068–1070.

2. I. Aleksander, "Array Networks for a Parallel Adder and Its Control," *IEEE Trans. Electr. Computers*, Vol. EC-16, No. 2, Apr. 1967.

3. S. F. Anderson et al., "The IBM System/360 Model 91: Floating-Point Execution Unit," *IBM Journal of R & D*, Vol. 11, No. 1, Jan. 1967, pp. 34–53.

4. J. M. Bratun et al., "Multiply/Divide Unit for a High-Performance Digital Computer," *IBM Tech. Disc. Bulletin*, Vol. 14, No. 6, Nov. 1971, pp. 1813–1316.

5. N. D. Kouvaras, et al., "Digital System of Simultaneous Addition of Several Binary Numbers," *IEEE Trans. Comp.*, Vol. C-17, No. 10, Oct. 1968, pp. 992–997.

6. O. L. MacSorley, "High-Speed Arithmetic in Binary Computers," *Proc. IRE*, Vol. 49, No. 1, Jan. 1961, pp. 67–91.

7. G. Metze and J. E. Robertson, "Elimination of Carry Propagation in Digital Computers," *Proc. International Conf. on Inf. Processing*, Paris, France, June 1959, pp. 389–396.

8. J. E. Robertson, "A Deterministic Procedure for the Design of Carry-Save Adders and Borrow-Save Subtractors," University of Illinois, Urbana-Champaign, Dept. of Computer Science, Report No. 235, July 1967.

9. F. A. Rohatsch, "A Study of Transformations Applicable to the Development of Limited Carry-Borrow Propagation Adders," Ph.D. Thesis, University of Illinois, Urbana-Champaign, June 1967.

10. S. Singh and R. Waxman, "Multiple Operand Addition and Multiplication, *IEEE Trans. Comp.*, Vol. C-22, no. 2, Feb. 1973, pp. 113–120.

11. J. Sklansky, "An Evaluation of Several Two-Summand Binary Adders," *IRE Trans. Elec. Comp.*, Vol. 9, No. 2, June 1960, pp. 213–225.

12. J. Sklansky, "Conditional-Sum Addition Logic," *IRE Trans. Elec. Comp.*, Vol. 9, No. 2, June 1960, pp. 226–231.

13. N. Takagi, H. Yasuura and S. Yajima, "High Speed VLSI Multiplication Algorithm with a Redundant Binary Addition Tree," *IEEE Trans. Comp.*, 34 (Sept. 1985) pp. 789–796.

14. C. S. Wallace, "A Suggestion for a Fast Multiplier," *IEEE Trans. Elec. Comp.*, Vol. EC- 13, Feb. 1964, pp. 14–17.

PROBLEMS

3.1 Design a conditional-sum adder to add 0011 with 1001. You may want to generate a table similar to the one in text first.

3.2 Perform $(101110)_2 + (100100)_2$ with a 6-bit conditional-sum adder and show how the result is obtained over the circuit. Estimate the total number of IC packages required assuming that triple CS cells are available in one IC package, as well as triple 2-input MUXs.

3.3 For the numerical example given below, fill the logic values in the parenthesis. Calculate the delay time for this example and for average.

3.4 (a) What is the delay time in the case of

$$(1010110110110110)_2 + (1001000110001110)_2$$

applying an asynchronous self-timing carry sensing adder?
(b) What is the delay time in the case of

$$(111100001111000011)_2 + (100011100000011011)_2?$$

(c) What is the average delay time over all the cases?
(d) What is the speed ratio of a 32-bit asynchronous self-timing carry sensing adder over the conventional ripple carry adder?

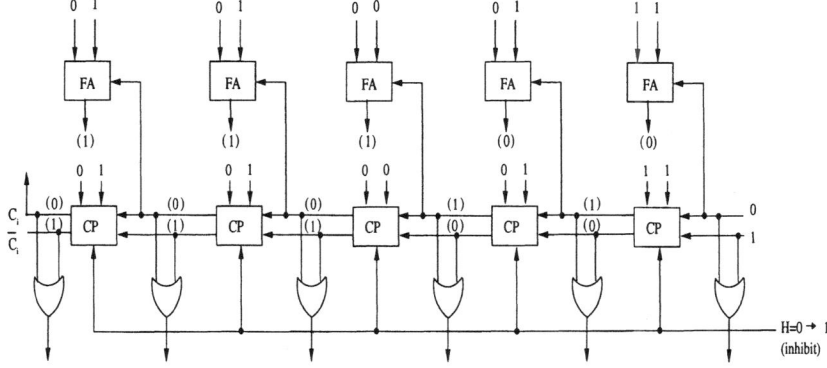

Fig. 3.11: Carry-Completion Sensing Adder

3.5 Suppose only 4-bit block carry lookahead (BCLA) and 4-bit carry lookahead (CLA) units are available. Construct a 64-bit adder/subtractor and estimate the delay time.
Hint: You may need three-levels of carry lookahead.

3.6 Given a 4-bit carry-save adder as follows, perform $3 + 5 + 1 + 6$. Fill in the results at the marked points after each clock cycle.

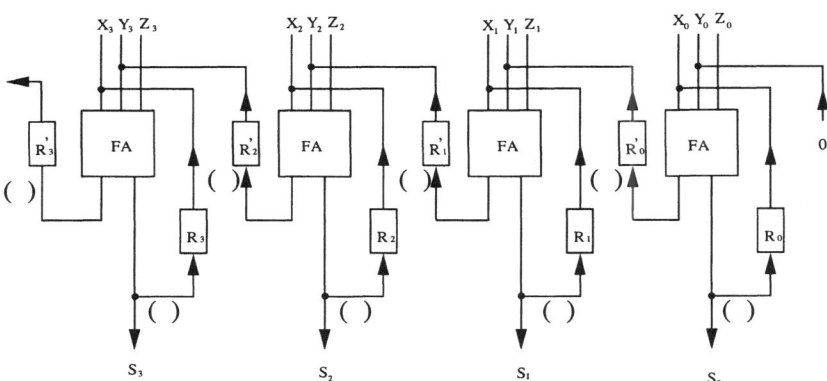

Fig. 3.12: Carry-Save Adder

3.7 (a) Design a carry save adder to add 12 numbers of 4-bit lengths. Find the delay time of it (in number of cycles).
(b) Construct a carry save adder tree of the same capability as in (a) with the minimum levels. Find the delay time and area complexity of it (in number of Δg and Ag).

3.8 How many full-adders and half-adders do you need to construct a bit-partitional adder which sums nine 3-bit numbers? Why?

3.9 (a) Design a bit-partitional adder to add 21 5-bit numbers.
(b) Based on the circuit from part(a), construct an adder to add 21 35-bit numbers.

3.10 Given the following 4-bit, 31-slice bit-partitional subadders (PSA), complete the design of a 16-bit 31-number adder with additional registers and CLA adders.

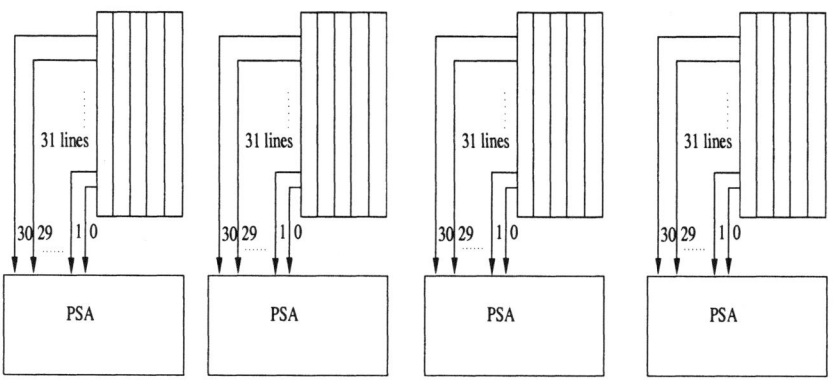

Fig. 3.13: Bit-Partitional Adder

4
Sequential Multiplication

Multiplication is an important task in computer arithmetic operations. Efficient algorithms and high-speed hardware should be developed to complete the multiplication. The sequential multiplication algorithms we introduce in this chapter are based on an add-shift approach. Detailed design of different types of multipliers will be given. Signed numbers multiplication will be discussed based on the indirect schemes and Robertson's approach. In addition, recoding techniques and Booth's algorithm, which speed up the multiplication by multiple-bit scanning, will be explained.

First we define some notation used in this chapter.

$\leftarrow$	replacement operator
$\circ$	concatenation or cascading operator
(R)	content of register R
$\vee$	logic OR operation
$\wedge$	logic AND operation
$+, -, \times, /$	arithmetic operations of addition, subtraction, multiplication and division.

Let the *multiplicand* A and the *multiplier* B be two n-bit unsigned numbers. The multiplication

$$P = A \times B$$

will create a $2n$-bit *product* P. For the case that A and B are signed numbers, $n-1$ bits are in each number excluding the sign bit. Hence $2(n-1) + 1 = 2n - 1$ bits will be in the product including the sign bit. To obtain a uniform representation, we add a dummy bit between the sign bit and the most significant bit, making a total of $2n$

78 SEQUENTIAL MULTIPLICATION

bits. The dummy bit can be either 0 or the replication of the sign bit in the product, depending on whether the signed numbers are in the sign-magnitude representation, or in 1's or 2's complement representation.

4.1 ADD-AND-SHIFT APPROACH

In the pencil-and-paper multiplication, the bits in the multiplier are examined one bit at a time. That bit multiplying the multiplicand results in a partial product, and a number of partial products are added up to form the product of the two given numbers. If the bit examined is 1, the multiplicand times 1 is performed and the multiplicand is the partial product. If the bit examined is 0, then zero is the partial product. For example,

$$
\begin{array}{r}
1\ 1\ 0\ 1 \\
\times\ )\quad 1\ 0\ 1\ 1 \\
\hline
1\ 1\ 0\ 1 \\
1\ 1\ 0\ 1 \\
0\ 0\ 0\ 0 \\
1\ 1\ 0\ 1 \\
\hline
1\ 0\ 0\ 0\ 1\ 1\ 1\ 1.
\end{array}
$$

The partial product can be generated by AND gates, since anything ANDed with 1 will be itself, and that ANDed with 0 will be zero. Instead of generating all the partial products first and adding them at the end, we add a partial product to the accumulated sum immediately after it is generated. Since each bit examined is one bit position higher than the prior one, each partial product to be added should be one bit position *left* than the prior one; or the partial product can be viewed steady while the accumulated sum is shifted *right* each time. Such a multiplication approach is hence called an add-and-shift approach.

A typical indirect multiply unit is composed of several functional devices as shown in Figure 4.1. Three registers such as *Accumulator (AC)*, *Multiplier Register (MR)* and *Auxiliary Register (AX)* are needed. Each has a length of n bits and the capability of parallel load. The signs of A and B are held in A_s and B_s, respectively. In the unsigned situation, zeros are loaded into these registers. An n-bit adder is incorporated to add the partial product to the intermediate result. A control counter (*CTR*) is required to keep track of the number of iterations carried and to notify the completion of the multiplication.

A typical indirect multiply procedure is presented as follows:

1. $MR \leftarrow B$;
   $AX \leftarrow A$;
   $AC \leftarrow 0$;
   $CTR \leftarrow 0$.

ADD-AND-SHIFT APPROACH 79

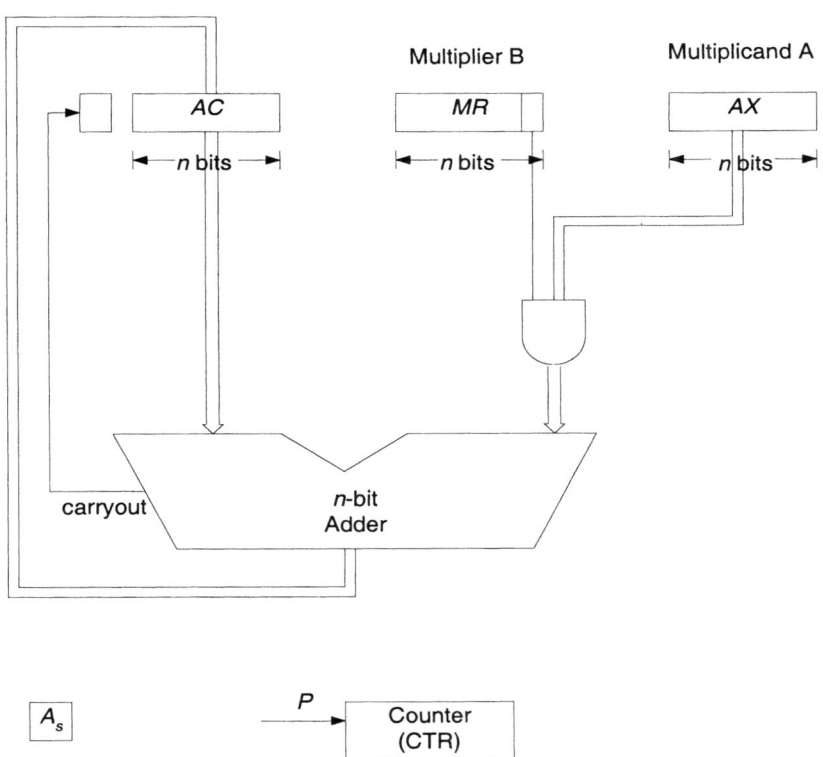

Fig. 4.1: Hardware for Sequential Multiplication

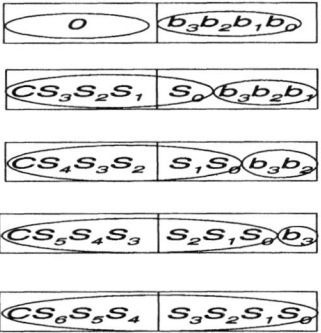

Fig. 4.2: Register Occupation

2. $AX \wedge MR_0$.

3. $C_{out} \cdot AC \leftarrow AC + (AX \wedge MR_0)$.

4. $AC \cdot MR \leftarrow C_{out} \cdot (AC_{n-1} \cdots AC_0) \cdot (MR_{n-1} \cdots MR_1)$.

5. $CTR \leftarrow CTR + 1$. If $CTR \neq n$, go to 2.

We give some marks to the above procedure. As an initiation, A is loaded into AX and B into MR. AC should be cleaned to make the intermediate result originally zero.

The least significant bit of MR, MR_0, is ANDed with the vector AX to obtain the partial product in Step 2. Depending on whether MR_0 is equal to 1 or 0, a copy of the multiplicand A or a zero vector will be sent to the adder.

In Step 3, the partial product obtained from Step 2 will be added to AC, in which the upper n-bit of the intermediate result obtained so far is stored. A carry out C_{out} may be generated during the addition.

Step 4 indicates a right shift operation of the sum in concatenating with the content in MR. The C_{out} is shifted into AC and MR_0 shifted out of MR.

Step 5 increments the counter and check whether n iterations have been executed so the procedure can halt.

This process needs to be repeated n times with n bits in the multiplier examined one by one. At the end of the procedure, the n-bit multiplier B will be pushed off the right end of MR, and the $2n$-bit product P should appear in the *cascaded register* $AC \cdot MR$. The sign of the result is in AC_{n-1} when a signed number multiplication is performed. For an example of 4-bit multiplication, the contents of AC and MR after each iteration are shown in Figure 4.2. We can see how the partial product grows in length and the multiplier fragment shrinks.

The delay time of the indirect multiply unit is

$$\triangle_T = \triangle_{SETUP} + n\triangle_{ADD-SHIFT} + \triangle_{STORE}.$$

Here,

$$\triangle_{ADD-SHIFT} = \triangle_{AND} + \triangle_{ADD} + \triangle_{LATCH},$$

and when n is large, the delay of add-shift becomes dominating.

4.2 INDIRECT MULTIPLICATION SCHEMES

We deal with the signed number multiplication in this section. Four standard schemes for indirect multiplication will be introduced: (1) unsigned number multiplication, (2) sign-magnitude number multiplication, (3) one's complement number multiplication and (4) two's complement number multiplication.

4.2.1 Unsigned Number Multiplication

The multiplication of two n-bit unsigned numbers is described by the flow chart in Figure 4.3. This operation is completed in $n + 1$ cycles. The setup such as loading data into appropriate registers requires one cycle, C_0. The multiply loop needs to be executed n times in cycle $C_1, \cdots\cdots, C_n$.

4.2.2 Sign-Magnitude Number Multiplication

In the multiplication of two sign-magnitude numbers, only $n - 1$ bits are in the magnitude of the multiplicand or multiplier. Hence $n - 1$ cycles are needed for the multiplication loop. The sign bits of A and B, a_{n-1} and b_{n-1}, are loaded into the flip-flops A_s and B_s in C_0, respectively, while the MSB of register AX and that of MR are initially loaded with zeros. The sign of the product can be decided at the end of the multiplication process by performing $A_s \oplus B_s$, and can be loaded into position AC_{n-1} while position AC_{n-2} remains 0.

The flow chart for the sign-magnitude number multiplication is given in Figure 4.4.

4.2.3 One's Complement Number Multiplication

From the above multiplication procedure, we can see that the magnitude of a product is obtained by multiplying the magnitudes of two input operands. If the input operands are represented in the complement form, those representing negative numbers should be converted to sign-magnitude form before entering the multiplication loop. Furthermore, the resulting product should be converted back to the complement form if it is negative. This kind of *pre-complement* and *post-complement* are included in the multiplication procedure, which is why the discussed multiplication method is referred to as "indirect multiplication".

For the 1's complement numbers, only bitwise negating is needed in the pre-complement or post-complement. So one additional cycle for each is sufficient.

82 SEQUENTIAL MULTIPLICATION

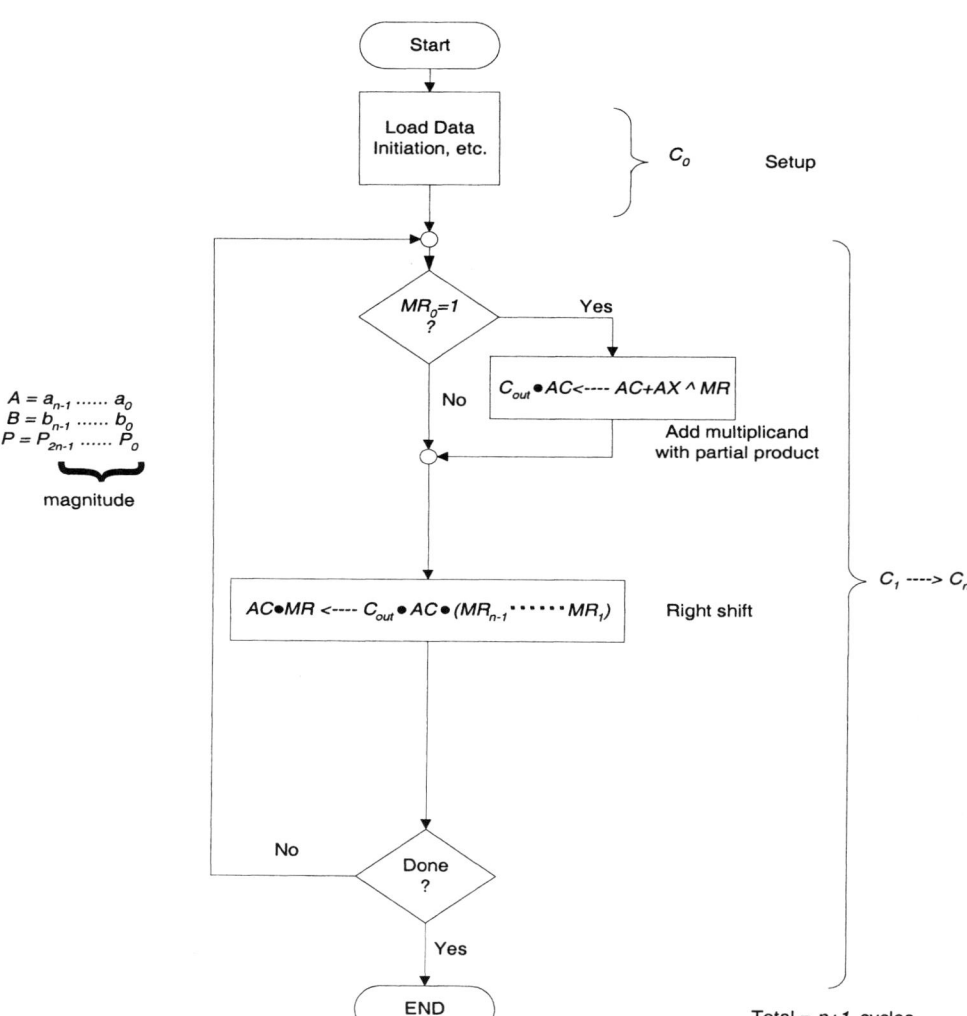

Fig. 4.3: Unsigned Number Multiplication

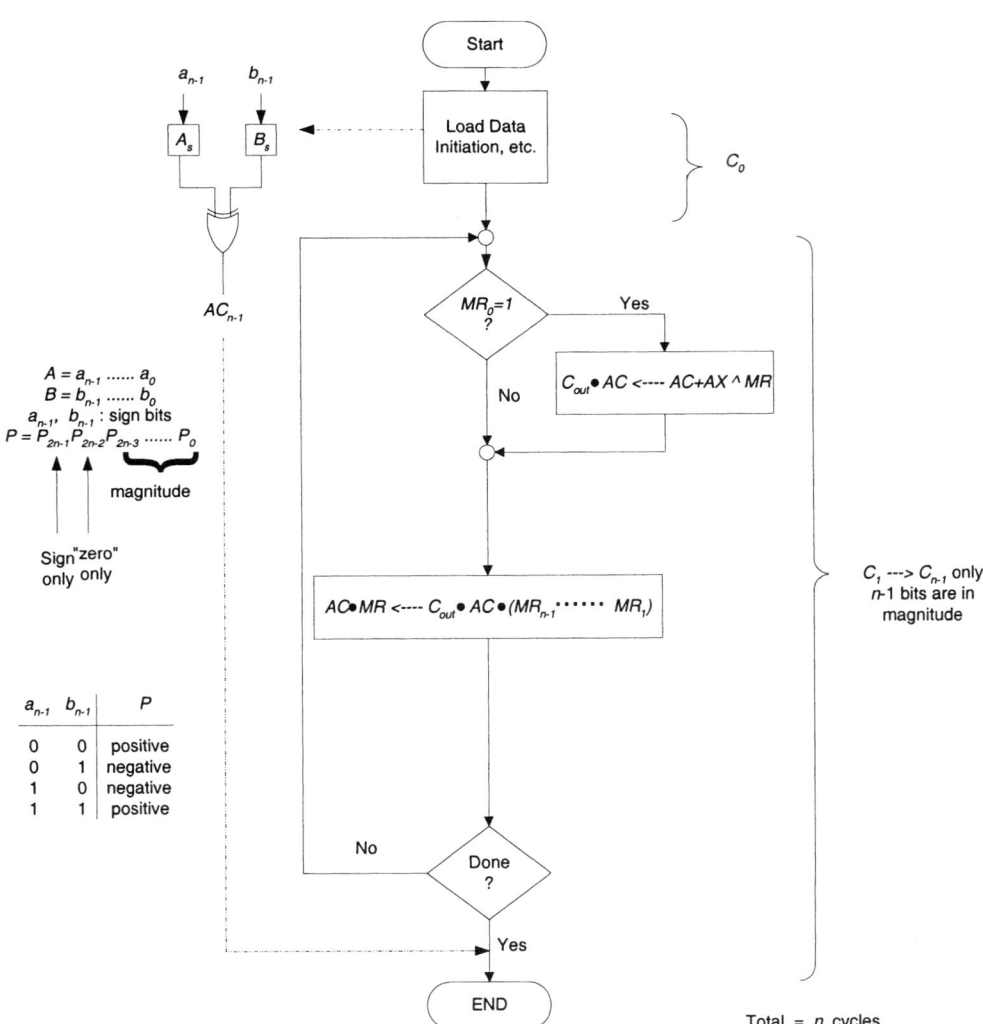

Fig. 4.4: Sign-Magnitude Number Multiplication

84 SEQUENTIAL MULTIPLICATION

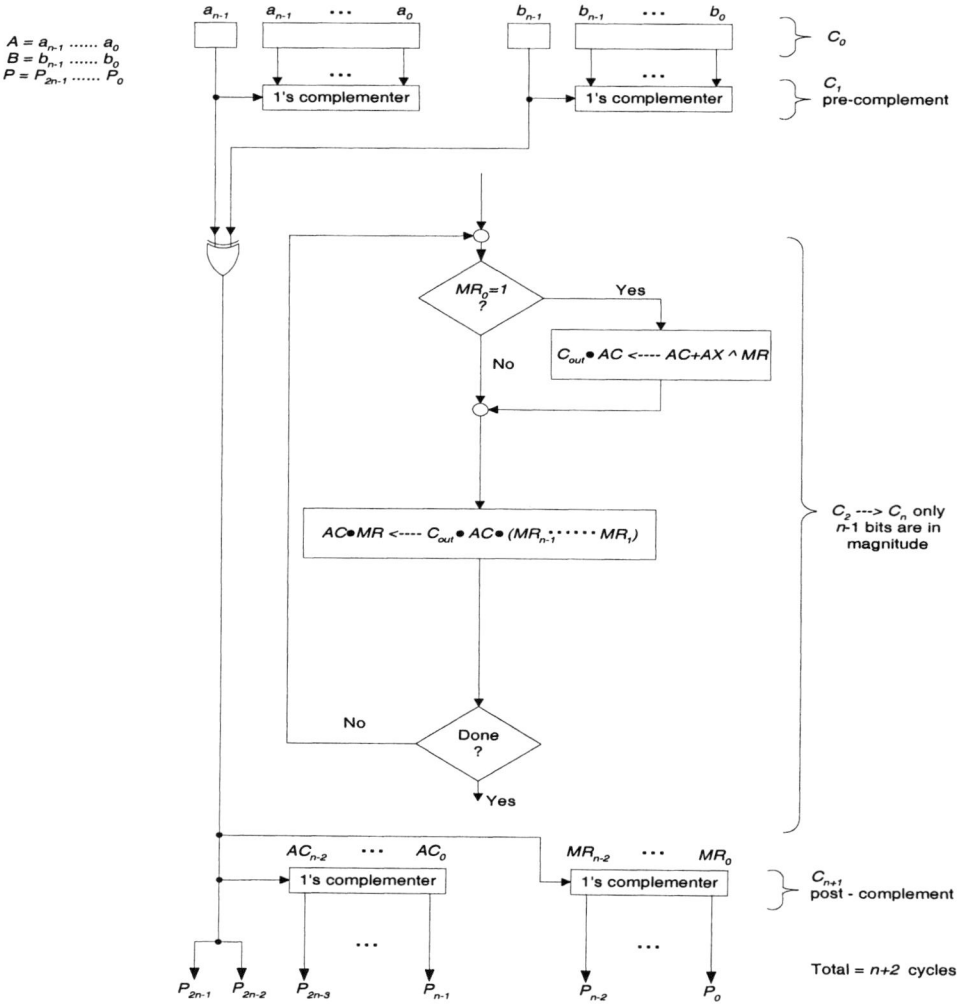

Fig. 4.5: One's Complement Number Multiplication

After the setup in C_0 and C_1 is dedicated to pre-complement, the $n-1$ cycles, from C_2 to C_n, are for the multiplication loop, and C_{n+1} is needed for post-complement. Figure 4.5 shows the flow chart for 1's-complement number multiplication. The total number of cycles required to multiply two n-bit 1's complement numbers is $n+2$.

4.2.4 Two's Complement Number Multiplication

The 2's complement operation is conducted by first performing the 1's complement operation and then adding a 1 to the LSB position. Two cycles are required for the pre-complement or post-complement. The pre-complement is completed in C_1 and C_2 after the setup cycle C_0. C_{n+2} and C_{n+3} are for the post-complement after executing the multiply loop in cycle C_3 to C_{n+1}.

In the post-complement operation, a $2n$-bit result is held in $AC \circ MR$ (see Figure 4.6), and the 2's complement of this long number is to be found. Here a "Zero MR" signal ZMR is generated by NORing all the n bits in register MR, and (1's complement + ZMR) is to be performed in AC. Recall that if the rightmost 1 can be identified in obtaining the 2's complement of a number, every bit on its left is simply to be negated.

In the case that the rightmost 1 is in MR, $ZMR = 0$. By performing (1's complement + 0) in AC, every bit there is negated. In the case that the rightmost "1" is in AC, $ZMR = 1$. Then (1's complement + 1), that is, 2's complement operation is to be performed in AC.

When all the bits in MR are 0s, we have all the 1s there after the 1's complement operation. To "+ 1", a carry will propagate from the LSB of MR to the AC, through all the 1s, since only the rightmost 0 (the rightmost 1 before the 1's complement operation) can stop the carry propagation and it is in AC now. By this mechanism, one can predict the 1 to be added on/carried into bit position n without wating for the carry propagation, hence reduce the length of carry propagation from $2n$ to n.

Recall that in the 2's complement operation, a bit is just negated if there is at least a 1 on the right of it. $ZMR = 0$ makes the "if" condition hold for every bit in AC.

$$AC \leftarrow (\overline{AC})$$

$$MR \leftarrow (\overline{MR}) + 1$$

are performed in this case where $\overline{AC}$ is the 1's complement of AC's content. $ZMR = 1$ when all the bits in MR are 0s. Computing $MR \leftarrow (\overline{MR}) + 1$ will result in the same MR and will generate a carry into AC in this case. So the following is to be performed:

$$MR = MR$$

$$AC \leftarrow (\overline{AC}) + 1,$$

where $ZMR = 1$ can be added to the LSB position of AC right away without waiting for the carry-in propagated from MR.

The complement operation is performed in cycle C_{n+2} and the increment in cycle C_{n+3}. $n + 4$ cycles in total are needed to perform an n-bit 2's-complement number multiplication. Presented in Figure 4.6 is a flow chart showing the procedure carried out.

The previously described algorithms can be combined to create a universal multiplier capable of performing any of the four types of multiplication.

86 SEQUENTIAL MULTIPLICATION

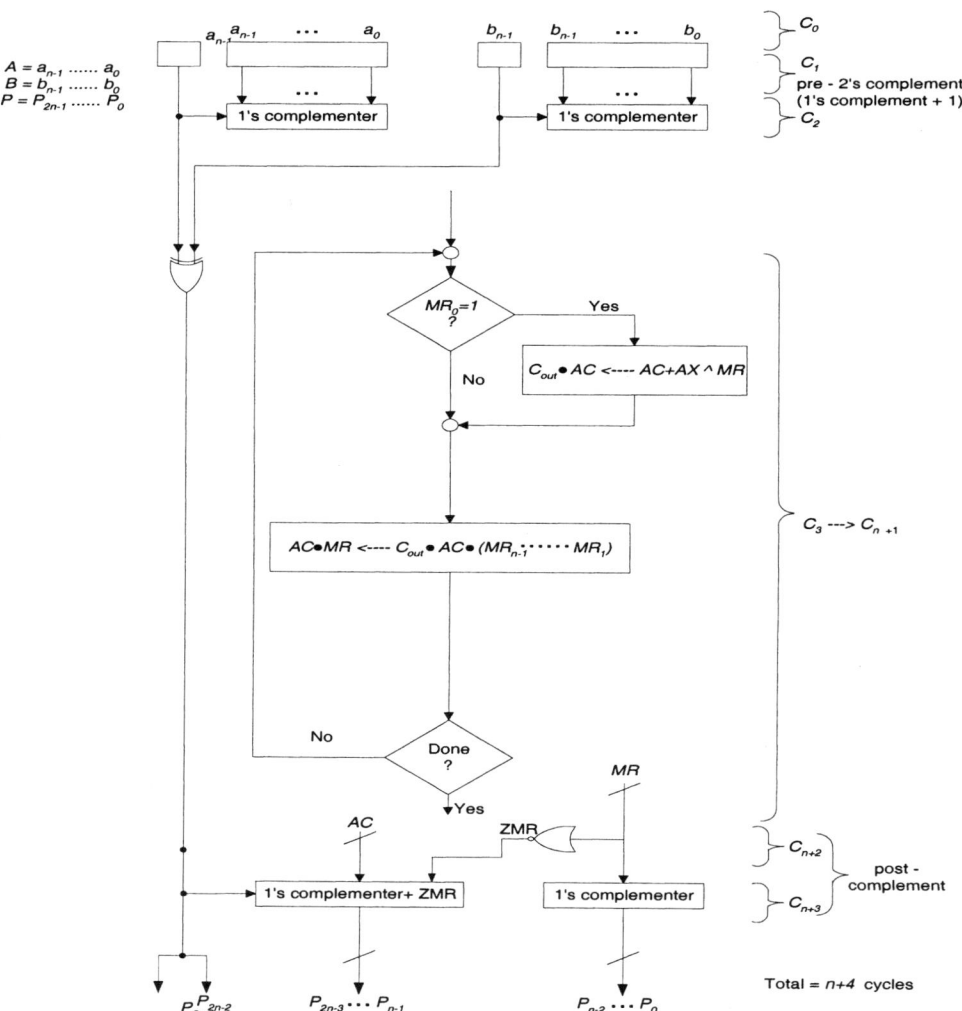

Fig. 4.6: Two's Complement Number Multiplication

$A = 1101_2 = -3_{10}$
$B = 0101_2 = 5_{10}$

			AC				MR			
Initially			0	0	0	0	0	1	0	1
$MR_0 = 1$		+)	1	1	0	1				
	Add A		1	1	0	1	0	1	0	1
	Shift		1	1	1	0	1	0	1	0
$MR_0 = 0$	Shift		1	1	1	1	0	1	0	1
$MR_0 = 1$		+)	1	1	0	1				
	Add A		1	1	0	0	0	1	0	1
	Shift		1	1	1	0	0	0	1	0
$MR_0 = 0$	Shift		1	1	1	1	0	0	0	1

$P = 11110001_2 = -15$

Fig. 4.7: Negative Multiplicand Times Positive Multiplier

4.3 ROBERTSON'S SIGNED NUMBER MULTIPLICATION

The pre- and post-complement presented in subsection 4.2.3 and 4.2.4 are time consuming though the idea is straightforward. A better procedure should be searched. Following are two cases discussed in the Robertson's Signed Number Multiplication.

Case 1: It has been suggested that if only the multiplicand is negative while the multiplier is positive, one can follow the procedure presented in Section 4.1. Even if the multiplicand is represented in the 2's complement form, it can be added to the accumulator without the pre-complement operation. However, when shifting the partial product to right, if the leftmost bit is 0 we shift in a 0. If the leftmost bit is 1 we shift in a 1. This is called *arithmetic shift*, in contrast to the logic shift which shifts in a 0 all the time. Figure 4.7 shows a numerical example of a 4-bit signed number multiplication for this case.

Case 2: If the multiplier is negative, a special correction step is needed in adopting the above procedure. That is, subtracting the multiplicand instead of adding it in the last step.

$A = 1101_2 = -3_{10}$
$B = 1011_2 = -5_{10}$

				AC				MR			
Initially				0	0	0	0	1	0	1	1
$MR_0 = 1$		+)		1	1	0	1				
	Add A			1	1	0	1	1	0	1	1
	Shift			1	1	1	0	1	1	0	1
$MR_0 = 1$		+)		1	1	0	1				
	Add A			1	0	1	1	1	1	0	1
	Shift			1	1	0	1	1	1	1	0
$MR_0 = 0$	Shift			1	1	1	0	1	1	1	1
$MR_0 = 1$		+)		0	0	1	1				
	Subtract A			0	0	0	1	1	1	1	1
	Shift			0	0	0	0	1	1	1	1

$P = 00001111_2 = 15$

Fig. 4.8: Negative Multiplicand Times Negative Multiplier

Let $B = 1b_{n-2}\cdots b_0$ be a negative multiplicand represented in the 2's complement form, with the sign bit $b_{n-1} = 1$. Then

$$\begin{aligned} |B| &= 2^n - B \\ &= 1(00\cdots 0) - (1b_{n-2}\cdots b_0) \\ &= 1(00\cdots 0) - (10\cdots 0) - (b_{n-2}\cdots b_0) \\ &= (10\cdots 0) - (b_{n-2}\cdots b_0) \\ &= 2^{n-1} - \sum_{0}^{n-2} b_i \cdot 2^i, \end{aligned}$$

hence

$$B = -|B| = -2^{n-1} + \sum_{0}^{n-2} b_i \cdot 2^i. \tag{4.1}$$

If we perform $A \times B$ following the procedure set for Case 1, the bit scan from b_0 to b_{n-2} is correct since it can be viewed as the second term in Equation (4.1) multiplying A. Bit b_{n-1} is 1 in the 2's complement representation of B, and when it is scanned an *add A* operation will be performed according to Case 1 procedure. Actually, $A \times 2^{n-1}$ is added due to the shifted bit position. Compared with the first term in Equation (4.1), $A \times 2^{n-1}$ should be subtracted rather than added. Here the last step in the Case 1 procedure should be corrected to subtract A, for the negative multiplier case. The example in Figure 4.8 shows the procedure designed for Case 2. Subtracting A is performed by adding the 2's complement of A.

4.4 RECODING TECHNIQUE

In general, the indirect multiplication schemes have a delay time proportional to n cycles where n is the length of the input operands. In order to speed up the multiplication, the multiple bit scanning approach will be adopted which can reduce the delay time with additional hardware.

4.4.1 Non-overlapped Multiple Bit Scanning

In the previously described multiplication schemes, one bit is scanned at a time. Depending on whether the LSB of *MR* is 1 or 0, a copy of the multiplicand A, and the intermediate result are added and then shifted, or just the intermediate result is shifted.

Now, instead of examining the LSB of *MR* only, two bits can be scanned at a time. In that case 0, A, $2A$ or $3A$ can be added to the intermediate result, depending on whether the least significant two bits in *MR* are 00, 01, 10 or 11. $2A$ is easy to obtain, just by shifting A left for one bit position. $3A$ can be obtained by $A + 2A$. Including the intermediate result, three operands are added up in this case. A multioperand adder is hereby requested. See Figure 4.9 for reference.

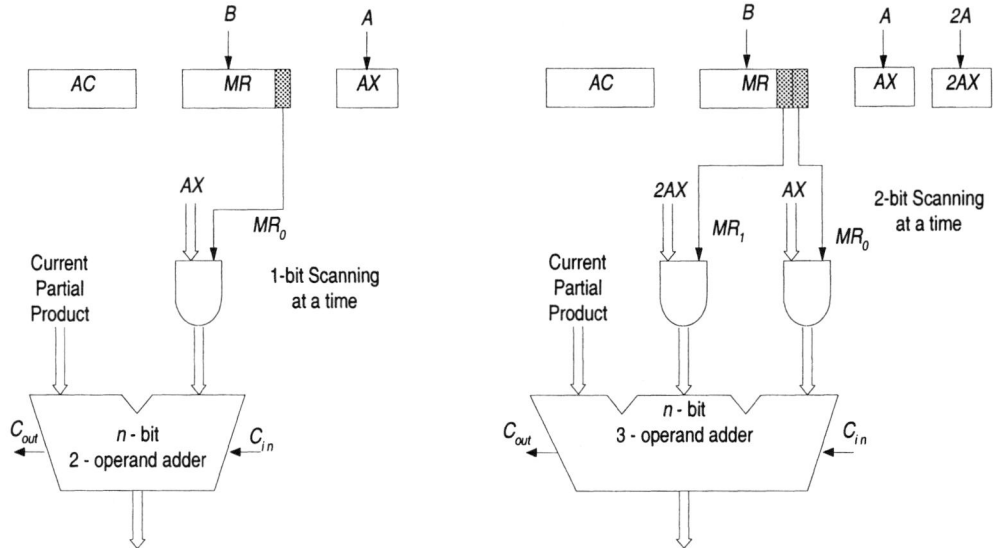

Fig. 4.9: Multiple Bit Scanning

Recall that when a single bit is scanned, the delay time is proportional to n cycles where each cycle time is related to the delay of a 2-operand adder. Now the delay time is proportional to $\frac{n}{2}$ cycles and each cycle time is related to the delay of 3-operand adder.

In general, for scanning m-bit at a time, the total delay time is proportional to $\frac{n}{m}$ cycles and each cycle time is related to the delay of $(m+1)$-operand adder.

4.4.2 Overlapped Multiple Bit Scanning

In the non-overlapped multiple-bit scanning algorithm, for each scan of the m-bit, m multiples of the multiplicand will be added to the intermediate result in the worst case. That is, $2^0 A + 2^1 A + 2^2 A + \cdots + 2^{m-1} A$. In the overlapped multiple-bit scanning algorithm to be introduced, we will see that only one multiple of A will be added.

We first look at the "string property" of a "conventional radix" integer number system.

$$2^{i+k} - 2^i = 2^{i+k-1} + 2^{i+k-2} + \cdots\cdots + 2^{i+1} + 2^i.$$

Figure 4.10 is an example to show how this property can be applied for recoding.

Let's call the consecutive 1s in the binary representation of a number *the string of 1s*, or just *string* for short. Suppose we scan the bits from right to left. Refer to the rightmost 1 in the string as the "beginning of string". For the leftmost 1, a 0 on its left is referred to as the "end of string". A long string of 1s with a 0 on the left

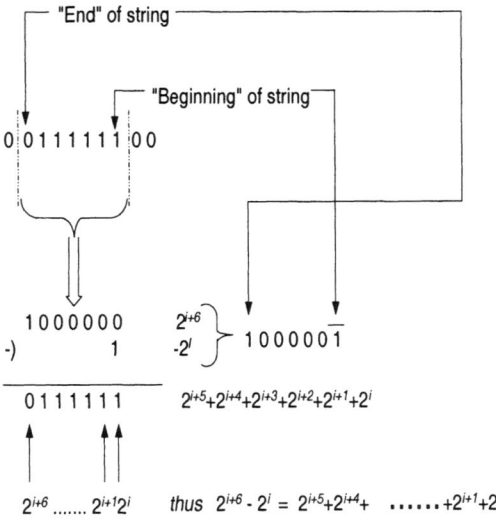

Fig. 4.10: String Property

can be recoded by putting a $\bar{1}$ in the position of "beginning of string", putting a 1 in the position of "end of string" and filling other positions with 0s. Here we use the notation in signed-digital representation $\bar{1}$ to refer to -1. That is,

$$0|0111111|00000|01111|0$$
$$= 0|100000\bar{1}|00000|1000\bar{1}|0.$$

In this way, when scanning the multiplier, say B, to perform a multiplication, we can replace the consecutive additions of multiplicand multiples by only one subtraction at the beginning of the string and one addition at the end of the string.

In the actual operation, the length of each scan is fixed. Suppose a few consecutive bits including their LSB in a scan are found 1s. It is unknown whether the LSB is in the middle of a string or at the beginning of the string. Overlapped scanning is hereby required. That is, let the LSB in the current scan overlap with the MSB in the previous scan. In other words, that bit is scanned twice. In the second scan it is used just as a reference bit, to tell the role of the bit on its left. In Table 4.1 each triplet is recoded and explained, and the corresponding operations are listed.

Figure 4.11 shows a 3-bit overlapped scan in contrast to the 2-bit non-overlapped scan of a 16-bit number. Note that a dummy 0 is attached to the right of the LSB so that if LSB is 1, the beginning of string (or isolated 1) case can be recognized. Also, the algorithm allows the scanning to take place from right to left, as well as from left to right. Moreover, each triplet could be scanned concurrently, which suggests a parallel scanning model.

Table 4.1: Recoding the Triplets

	2^2	2^1	2^0			
		Scanned bits				
	X_{i+3}	X_{i+2}	X_{i+1}	X_i	X_{i-1}	Multiples of 0, 2A or 4A	String position
	d	0	0	0	d	$\Rightarrow 0$	No string of "1"
	d	0	0	1	d		
=	d	0	1	0	d	$\Rightarrow +2A$	End of string
	d	0	1	0	d	$\Rightarrow +2A$	Isolated "1"
	d	0	1	1	d		
=	d	1	0	0	d	$\Rightarrow +4A$	End of string
	d	1	0	0	d		
=	d	$\bar{1}$	0	0	d	$\Rightarrow -4A$	Beginning of string
	d	1	0	1	d		
=	d	$\bar{1}$	1	0	d	$\Rightarrow -2A$	Beginning and end of string
	d	1	1	0	d		
=	d	0	$\bar{1}$	0	d	$\Rightarrow -2A$	Beginning of string
	d	1	1	1	d		
=	d	0	0	0	d	$\Rightarrow 0$	Middle of string

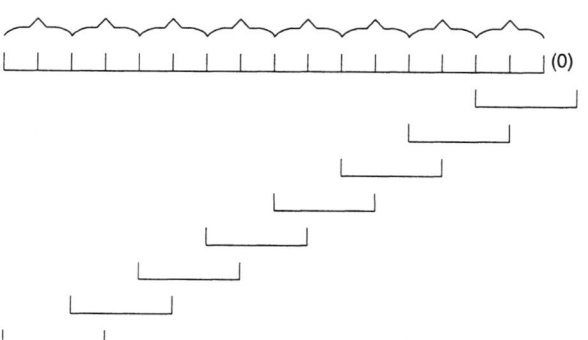

Fig. 4.11: Two-Bit Scan vs. Overlapped Three-Bit Scan

4.4.3 Booth's Algorithm

Based on the string property described above, we can recode the multiplier B to D by finding for two bits $b_i b_{i-1}$ a bit d_i in the following scheme.

b_i	b_{i-1}	d_i	
0	0	0	no string
0	1	1	end of string
1	0	$\bar{1}$	beginning of string
1	1	0	center of string

Booth's Algorithm has the following operations designed accordingly:

$$d_i = \begin{cases} 0, & \text{if } b_i = b_{i-1} \\ 1, & \text{if } b_i < b_{i-1} \\ \bar{1}, & \text{if } b_i > b_{i-1} \end{cases}$$

Recall that the beginning of the string is identified by a 0 on the right of the consecutive 1s in the recoding table.

A dummy 0 is attached to the right of the LSB. The scan can be performed from right to left, from left to right or in parallel, and the result should be the same. Given

$$A = a_{n-1} \cdots a_1 a_0 \quad multiplicand$$

and

$$B = b_{n-1} \cdots b_1 b_0 \quad multiplier,$$

the Booth's Algorithm can be described as follows:

```
Initialize b_-1 = 0, and i = 0.
Add A to partial product and shift right if b_i < b_{i-1},
Add -A to partial product and shift right if b_i > b_{i-1},
Shift right if b_i = b_{i-1}.
Increment i, go to 2 if i < n.
```

Here the Booth's multiplication can multiply two 2's complement numbers directly without special care of the signs of operands. In other words, pre-complement and post-complement are not needed and direct multiplication can be performed. Note that the shift here is an arithmetic shift. That is, if the MSB (sign bit) is 1, shift in a 1, if it is 0, then shift in a 0.

An example performing $5 \times (-6)$ applying Booth's Algorithm is presented in Figure 4.12. Considering that $P = 5 \times (-6) = -30$, we can find the result of the above procedure correct.

$A = 5 = 0101$ $\qquad\qquad\qquad -A = 1011$
$B = -6 = 1010$ $\qquad\qquad P = 5 \times (-6) = -30$

		AC	MR	B_{-1}			
$i=0$		0000	1010	0		$b_i = b_{i-1}$	Shift right
$i=1$	"0" →	0000	0101	0	0→	$b_i > b_{i-1}$	Add $-A$ and shift
	+	1011					

		1011	0101	0			
$i=2$	"1"→	1101	1010	1	0→	$b_i < b_{i-1}$	Add A and shift
	+	0101					

		0010	1010	1			
$i=3$	"0"→	0001	0101	0	1→	$b_i > b_{i-1}$	Add $-A$ and shift
	+	1011					

		1100	0101	0
	"1" →	1110	0010	

$P = 11100010_2 = -30$
$(00011110_2 = 16 + 8 + 4 + 2 = 30)$

Fig. 4.12: Example of Booth's Multiplication

Notice that the most significant two bits both represent the sign of the product, and $(11100010)_2$ is the 2's complement representation of -30. This method is attractive for such B that contains long subsequences of 1s. It is not so efficient when many isolated 1s are contained, since after recoding D may contain more non-zero bits than the original B. To overcome this drawback, higher radix string recoding has been investigated.

Rather than the above table exploring the radix-2 string recoding, a radix-4 string recoding can be carried on in a similar way. The input is of three bits, b_i, b_{i-1} and b_{i-2}, and the output is of 2 bits, d_i and d_{i-1}.

b_i	b_{i-1}	b_{i-2}	d_i	d_{i-1}
0	0	0	0	0
0	0	1	0	1
0	1	0	0	1
0	1	1	1	0
1	0	0	$\bar{1}$	0
1	0	1	$\bar{1}$	1
1	1	0	0	$\bar{1}$
1	1	1	0	0

4.4.4 Canonical Multiplier Recoding

This recoding technique uses signed-digit (SD) representation. Recall that a *minimal* SD vector has a minimal number of non-zero digits.

For a special class of the minimal SD vectors, we give the definition of the *canonical signed-digit vector* below. The minimal SD vector $D = d_{n-1} \cdots d_1 d_0$ that contains no adjacent nonzero digits is called a canonical signed-digit vector. That is, $d_i \times d_{i-1} = 0$ for $1 \leq i \leq n-1$.

In the representations listed above, $(00010\bar{1})_2$ is a canonical SD vector, whereas $(000011)_2$ is not.

It has been prove that any $D = d_{n-1} d_{n-2} \cdots d_1 d_0$ with a fixed value and a fixed length can be transformed into a "unique" canonical SD form if $d_{n-1} \times d_{n-2} \neq 1$. By imposing an nth digit $d_n \equiv 0$, we can always achieve this. In effect, we will always assume that $D = d_n d_{n-1} \cdots d_1 d_0$ has $n + 1$ bits and $d_n \equiv 0$.

Once we obtain the canonical SD form of a number, say the multiplier B which we are interested in, we know that this canonical form has $\lceil \frac{n+1}{2} \rceil$ non-zero bits or digits. In the design of a multiplier then, only the addition of A or $(-A)$ should be facilitated. Note that the 0s cause shifts rather than additions, and the delay time $\Delta_{shift} \ll \Delta_{add}$.

Next, let's look at how a given $(n + 1)$-bit number, $(0b_{n-1} \cdots b_0)$, can be transformed into an $(n + 1)$-bit canonical SD number, $D = d_n d_{n-1} \cdots d_0$, (note that $b_n \times b_{n-1} = 0$,) such that $\sum_{i=0}^{n} b_i \times 2^i = \sum_{i=0}^{n} d_i \times 2^i$. The algorithm can be described as follows.

96 SEQUENTIAL MULTIPLICATION

```
Start with LSB, b₀, of B and let i = 0 and c₀ = 0.
Find c_{i+1} = b_{i+1} ∧ b_i ∨ b_i ∧ c_i ∨ b_{i+1} ∧ c_i.
Find d_i = b_i + c_i - 2c_{i+1}.
i = i + 1. If i = n, stop; else, go to step 2.

Stop: c_n d_{n-1} d_{n-2} ··· d_0 is the (n+1)-bit canonical SD.
```

Based on Step 2 and 3, a table is generated as follows.

Conventional Multiplier Bits		Assumed Carry-in	Recoded Bit	Carry-out
b_{i+1}	b_i	c_i	c_{i+1}	d_i
0	0	0	0	0
0	0	1	0	1
0	1	0	0	1
0	1	1	1	0
1	0	0	0	0
1	0	1	1	$\bar{1}$
1	1	0	1	$\bar{1}$
1	1	1	1	0

Let's work on a numerical example with $B = (0\ 1\ 0\ 1\ 0\ 1\ 1)_2$.

$$
\begin{array}{llllll}
i = 0 & b_1 b_0 = 11 & c_0 = 0 & \Rightarrow & c_1 = 1 & d_0 = \bar{1} \\
i = 1 & b_2 b_1 = 01 & c_1 = 1 & \Rightarrow & c_2 = 1 & d_1 = 0 \\
i = 2 & b_3 b_2 = 10 & c_2 = 1 & \Rightarrow & c_3 = 1 & d_2 = \bar{1} \\
i = 3 & b_4 b_3 = 01 & c_3 = 1 & \Rightarrow & c_4 = 1 & d_3 = 0 \\
i = 4 & b_5 b_4 = 10 & c_4 = 1 & \Rightarrow & c_5 = 1 & d_4 = \bar{1} \\
i = 5 & b_6 b_5 = 01 & c_5 = 1 & \Rightarrow & c_6 = 1 & d_5 = 0 \\
i = 6 & \text{Stop.}
\end{array}
$$

So, we have $D = (1\ 0\ \bar{1}\ 0\ \bar{1}\ 0\ \bar{1})$.

An example is given in Figure 4.13(a) showing the scanning pattern in a 32-bit multiplication using the generalized Booth's Algorithm. A dummy bit is attached to the LSB of the multiplier, and a total of 33 bits are to be scanned. Eight iterations of the scan are conducted sequentially, each inspecting 5 bits with one bit overlapped.

The first iteration scans 5 bits. The other seven iterations scan 28 new bits with 4 new bits per iteration excluding the overlapped bit. Within one iteration, 5 bits are taken care of by the overlapped 2-bit scanning applying the Booth's Algorithm.

The multiplicand multiples of A, $-A$ and 0 can be first prepared. The $-A$ is obtained by adding a "1" to the 1's complement of A. Then for each pair of bits scanned, one of the multiples will be selected depending on the value of the 2 bits $b_i b_{i-1}$. Figure 4.13(b) gives the schematic of logic circuit selecting the appropriate multiplicand multiples. Gate 2 will be blocked by the output of Gate 4 when $b_i = b_{i-1}$, and its output will be 0 which is the multiple that should be selected. When $b_i > b_{i-1}$, A will be negated after passing Gate 1, and become the 1's complement of A. It will

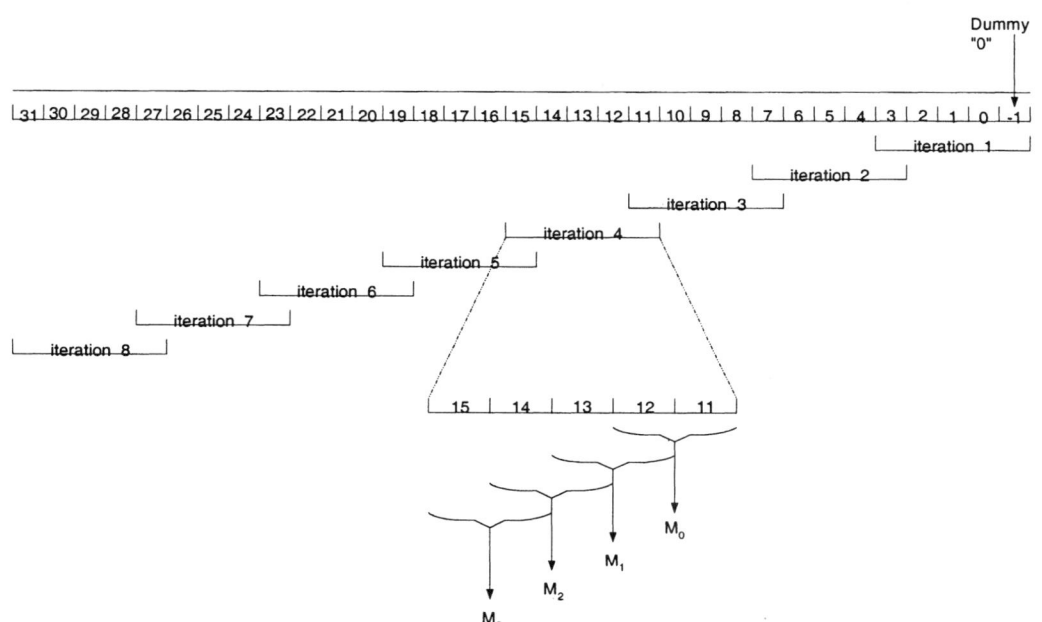

(a) Scan Pattern

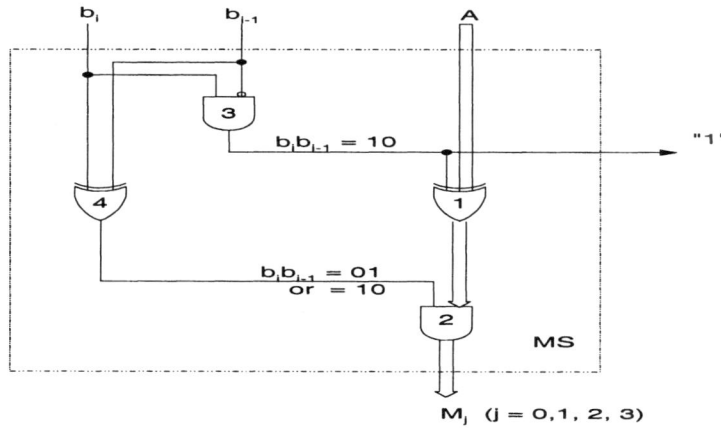

(b) Multiples Selection

Fig. 4.13: Scan Pattern in 32-bit Multiplication

be routed through Gate 2, and a "1" is also output from Gate 3. They will be added up in the future addition to form the 2's complement of A, $-A$. The other case to make Gate 2 open is $b_i < b_{i-1}$. A will remain unchanged after passing Gate 1 in this case, and the multiple to be selected is A.

The four multiples, M_0, M_1, M_2 and M_3 can be obtained from the four bit-pair scans in parallel with the replicated MS cells. Then they are to be added up. It should be noticed that within the same iteration, one multiple is off a bit position than its adjacent. A 6-input CSA tree is applied to trade for the operating speed with more expensive hardware (see Figure 4.14). The saved carry and sum are fed back to the input side, and the carry vector is shifted one bit left than the sum vector.

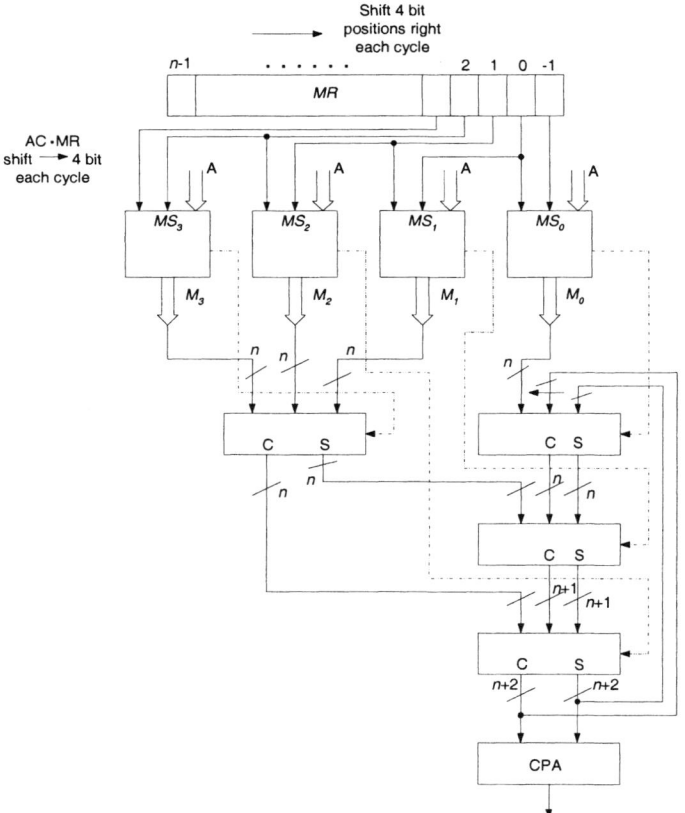

Fig. 4.14: Adding the Bit-Pairs Parallelly Scanned with a CSA Tree

One iteration after another, the group of four multiples are continuously generated one group after another. Each group is added to the accumulative partial product by the CSA tree, and the addition of different groups can be pipelined. Two outputs, sum and carry are generated by the CSA tree. To get the final result they need to be added by a two-operand fast adder. However, if new addends come in and the addition repeats, the carry can be saved and the two outputs can be fed back to the input of the CSA tree to be added with new inputs.

To perform an n-bit multiplication applying such 5-bit per iteration scan pattern where 4 new bits are examined at a time excluding the overlapped bit, $\lceil \frac{n+1}{4} \rceil$ iterations are required. With the simplest hardware, on the other hand, it takes n cycles to scan all the n bits.

REFERENCES

1. S. F. Anderson, et. al. "The IBM System/360 Model 91: Floating-Point Execution Unit," *IBM Journal of Research and Development*, Jan. 1967, pp. 34–53.

2. A. Avizienis, "Recoding of the Multiplier," Class Notes, Engr. 225A, UCLA, Los Angeles, CA, 1971.

3. A. D. Booth, "A Signed Binary Multiplication Technique," *Quart. Journ. Mech. and Appl. Math.*, Vol. 4, Part 2, 1951, pp. 236–240.

4. V. S. Burstev, "Accelerating Multiplication and Division Operations in High-Speed Digital Computers," Tech. Report, Institute of Exact Mechanics and Computing Technique, The Academy of Sciences of the USSR, Moscow, 1958.

5. P. M. Fenwick, "Binary Multiplication with Overlapped Addition Cycles," *IEEE Trans. Comput.*, Vol. C-18, No. 1, Jan. 1969, pp. 71–74.

6. H. Freeman, "Calculation of Mean Shift for a Binary Muliplier Using 2, 3, or 4-bit at a Time," *IEEE Trans.*, Vol. EC-16, No. 6, Dec. 1967, pp. 864–866.

7. A. A. Kamal and M. Ghanam, "High-Speed Multiplication System," *IEEE Trans. Comp.*, Vol. C-21, No. 9, Sept. 1972, pp. 1017–1021.

8. M. Lehman, "Short-Cut Muliplication and Division in Automatic Binary Digital Computers," *Proc. IEEE*, Vol. 10, Sept. 1958, pp. 496–504.

9. H. Ling, "High-Speed Computer Multiplication Using a Multiple-Bit Decoding Algorithm," *IEEE Trans. Comp.*, Vol. C-19, No. 8, Aug. 1970, pp. 706–709.

10. O. L. MacSorley, "High-Speed Arithmetic in Binary Computers," *Proc. IRE*, Vol. 49, Jan. 1961, pp. 91–103.

11. G. Metze, "A Study of Parallel One's Complement Arithmetic Units with Separate Carry or Borrow Storage," Ph.D. Thesis, Univ. of Illinois, Urbana, 1958.

12. F. J. Mowle, "Simplified Logic Design Using Digital Circuit Elements," Vol. 3, Class Notes, Dept. of Elec. Eng., Purdue University, West Lafayette, IN, Sept. 1974, Chap. 14.

13. J. O. Pennhollow, "Study of Arithmetic Recoding with Applications in Multiplication and Division," Ph.D. Thesis, Univ. of Illinois, Urbana, Sept. 1962.

14. G. W. Reitwiesner, "Binary Arithmetic," in *Advances in Computers*, Vol. 1, Academic Press, New York 1960, pp. 261–265.

15. J. E. Robertson, "Two's Complement Multiplication in Binary Parallel Digital Computers," *IRE Trans.*, Vol. EC-4, No. 3, Sept. 1955, pp. 118–119.

16. J. E. Robertson, "The Correspondence Between Methods of Digital Division and Multiplier Recoding Procedures," *IEEE Trans. Comp.*, Vol. C-19, No. 8, Aug. 1970, pp. 692–701.

17. K. S. Trivedi and M. D. Ercegovac, "On-Line Algorithms for Division and Multiplication," *IEEE Trans. Comp.*, Vol. C-26, No. 9, July 1977, pp. 681–687.

18. C. S. Wallace, "A Suggestion for a Fast Multiplier," *IEEE Trans. Electronic Comp.*, Vol. EC-14, No. 1, Feb. 1964, pp. 14–17.

PROBLEMS

4.1 Use the add-and-shift approach to perform the unsigned number multiplication of $(11001)_2 \times (10110)_2$.
(a) Show the typical sequential multiplication procedure, referring to the procedure in Section 4.1.
(b) Build a table showing the register occupation during this procedure.
(c) Find the delay time of this sequential multiplication.

4.2 Consider the multiplication of two numbers, $A=1010$ and $B=1101$, for two cases:
(a) They are both unsigned numbers;
(b) They are both signed numbers.
Draw a flow chart for each case.

4.3 Using 2's complement number multiplication to perform the multiplication in Problem 2. Show the corresponding flow chart.

4.4 Let $A = -7$, $B_1 = 9$ and $B_2 = -9$.
(a) What are the 2's complement expressions of A, B_1 and B_2?
(b) Perform $A \times B_1$ and $A \times B_2$ using Robinson's method, and compose a table showing the procedure of each multiplication.

4.5 Given the hardware as in Figure 4.9(b), perform the unsigned number multiplication $(100100)_2 \times (100111)_2$, where $(100100)_2$ is the multiplicand. Show the performed operations and the content of $AC \circ MR$ register step by step if non-overlapped 2-bit scanning is applied.

4.6 Given a 4-bit adder and two 4-bit registers AC and MR, perform $P = A \times B$ by Booth's Algorithm, where $A = (0101)_2$ and $B = (1001)_2$ are two unsigned numbers.

4.7 Given $A = 13$ and $B = -10$, show the procedure of the multiplication of A and B applying Booth's Algorithm.

4.8 Build a table for an overlapped 4-bit scan. Find for each case the multiples of the multiplicand and the string property your work is based on.

4.9 Given a 9-bit signed number $X = (100110111)_2$, if using an overlapped 3-bit scan to recode X, show how the multiples of multiplicand A resulted in different scans. How do they add to the final answer which is the X multiples of A?

4.10 The overlapped scanning multiplication algorithm allows the scan to take place from right to left or left to right. For the number given in Problem 4.9, verify that the correct answer can be obtained by scanning X from left to right.

4.11 Given a 32-bit number for scan, an overlapped 8-bit scan is to be applied in sequence. How many iterations are needed? For each iteration in the sequential scans, Booth's Algorithm is to be applied in parallel to recode the scanned 8 bits. How many multiples of the multiplicand will be added at a time by an adder? Draw a chart to show your design of the scan pattern.

4.12 Given a 16-bit binary number $B = (0011101101110011)_2$, find the corresponding canonical signed-digit vector D which represents the same value as B. Show step-by-step how each digit in D is obtained.

5
Parallel Multiplication

In the preceding chapter, a multiplication function was completed in n iterations given the input operand of length n. In each iteration, basically an add-and-shift operation was performed. In this chapter, instead of repetitively using the same single-stage hardware, an Array Multiplier is proposed, in which multiple stages of hardware are provided by replicating the single stage circuit. The different iterations of addition can now be performed by different stages of hardware, and each shifting is realized by the proper wiring between hardware stages. In this case, multiple multiplication operations can be overlapped. That is, the second multiplication can start before the first one completes, hence the time required to complete multiple multiplication functions can be reduced. To date, various iterative array multipliers and cellular array processors are available that are capable of high-speed multiplication demanded in scientific computation.

5.1 WALLACE TREES

In the multiplication operation, the addition of the partial products is the most time-consuming process. In the previous chapter, how to reduce the number of partial products is discussed. Here we present a hardware to handle the repetitive addition – Wallace tree.

A *Wallace tree* is a bit-slice adder which adds all the bits in the same bit position. Figure 5.1 gives the Wallace trees for different number of inputs. In (a), a 3 bit-slice adder, W_3, is presented which is actually a 3-input 2-output carry-save full adder.

104 PARALLEL MULTIPLICATION

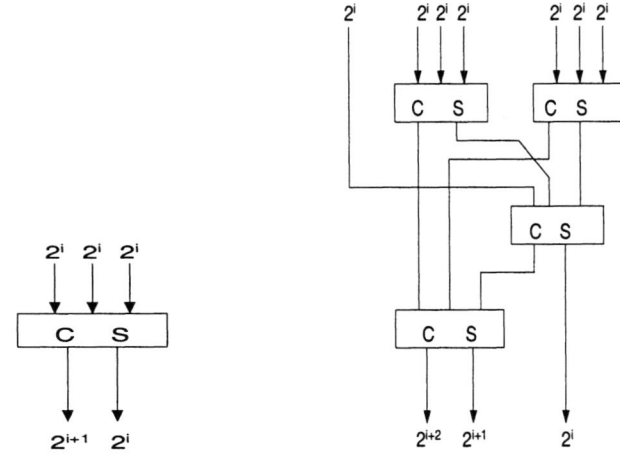

(a) 3-Bit Wallace Tree

(b) 7-bit Wallace Tree

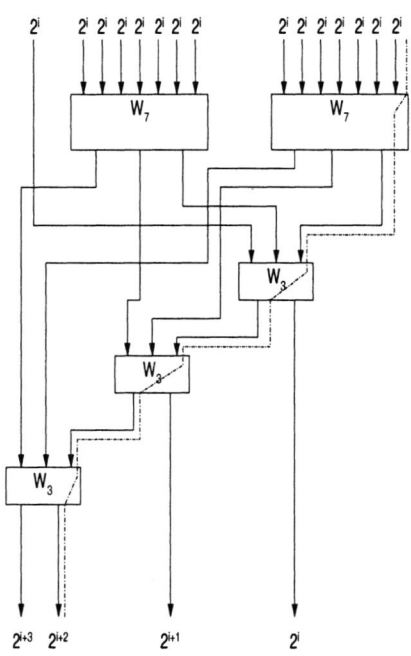

(c) 15-bit Wallace Tree

Fig. 5.1: Wallace Tree

Table 5.1: Combination and Delay of k-input Wallace Tree

k	W_i **Combination**	$\triangle_T$
3	W_3	$\triangle_{FA} = 2\triangle_g$
6	$4 \times W_3$	$3\triangle_{FA} = 6\triangle_g$
7	$(4 \times W_3)$ or (W_7)	$3\triangle_{FA} = 6\triangle_g$
15	$(11 \times W_3)$ or $(3 \times W_3, 2 \times W_7)$	$5\triangle_{FA} = 10\triangle_g$

Figure 5.1(b) shows a 7-input Wallace tree, W_7, which is formed by four W_3s. Note that the delay time along the critical path is $3\triangle_{FA}$. After $2\triangle_{FA}$ time, the output in the 2^i bit position can be ready.

A 15 bit-slice Wallace tree, W_{15}, is also given, which is composed of two W_7s and three W_3s. Along the critical path (marked by the dashed line), the delay time is

$$\triangle_T = 2\triangle_{FA} + \triangle_{FA} + \triangle_{FA} + \triangle_{FA}$$
$$= 5\triangle_{FA}.$$

In general, a k-input Wallace tree is a bit-slice summing circuit which produces the sum of k bits all having equal weights. Table 5.1 lists the combination and delay time for typical k-input Wallace trees. Note that a k-input Wallace tree has $\lceil log_2(k+1) \rceil$ outputs.

5.2 UNSIGNED ARRAY MULTIPLIER

In this section we introduce an unsigned array multiplier. Given $A = (a_{m-1} \cdots a_1 a_0)_2$ and $B = (b_{n-1} \cdots b_1 b_0)_2$, where A and B are unsigned integers, suppose $P = (p_{m+n-1} p_{m+n-2} \cdots p_1 p_0)_2$ is the product. We have

$$\begin{aligned} P &= A \times B \\ &= \sum_{i=0}^{m-1} a_i \cdot 2^i \times \sum_{j=0}^{n-1} b_j \cdot 2^j \\ &= \sum_{i=0}^{m-1} \sum_{j=0}^{n-1} (a_i b_j) \cdot 2^{i+j} \\ &= \sum_{k=0}^{m+n-1} p_k \cdot 2^k. \end{aligned}$$

Note that A is of m bits and B n bits, and the product P has $m + n$ bits. Such multiplication is referred to as m-by-n multiplication. If A and B are of the same length, each containing n bits, then P is $2n$-bit long. We refer to such multiplication

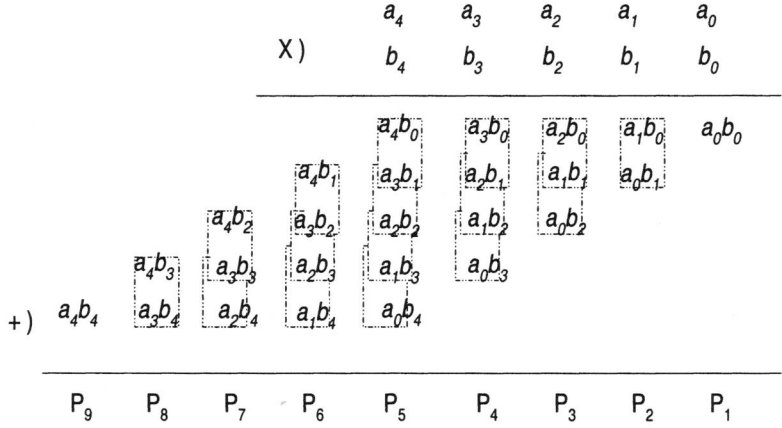

Fig. 5.2: 5-by-5 Multiplication

as n-by-n multiplication, or just n-bit multiplication. Figure 5.2 shows a 5-by-5 multiplication.

Each $a_i b_j$, $0 \leq i, j \leq n-1$, is called a *summand*, and each row in the middle section is called a partial product. Two kinds of operations are required here: the generation of summands and the addition of summands. The summands can be generated in parallel provided that an array of $m \times n$ AND gates is available. The summand summation can be completed by an array of $(m - 1) \times n$ FAs. The summands in a box in Figure 5.2 can be added by an full-adder (FA). If there is a box above it, the summand is added to the sum generated by that box. Figure 5.3 shows a 5-by-5 unsigned number multiplication performed by a 4×5 array of FAs. In row j, an input vector $(a_4 b_{j+1}, a_3 b_{j+1}, a_2 b_{j+1}, a_1 b_{j+1}, a_0 b_{j+1})$ is input from the right of the FAs. Two vectors are produced, a carry vector and a sum vector. The carry is connected one position to the left in the succeeding row where it is added with the input vector from the right and the sum vector from the above row. In the top row, only the input vector is added with $(a_4 b_0, a_3 b_0, a_2 b_0, a_1 b_0, a_0 b_0)$, so a row of half-adders (HAs) can be used if unity is not a concern. In the bottom row, the carry is rippled from right to left. One can see from Figure 5.2 that $p_0 = a_0 b_0$ and no addition is needed in the LSB position. That is why a 4×5 array is sufficient to perform the 5-by-5 multiplication rather than a 5×5 array.

The delay time of summands generation is the same as that of AND gate. Hence,

$$\triangle_{sg} = \triangle_{AND}.$$

For the delay time of summands summation, not all the bits p_j in the product form at the same time. The input values must be maintained long enough to allow all the outputs to become stable. The worst case propagation delay path should be identified here. Starting from the right most cell of the top row, there is a delay time of $4\triangle_{FA}$ before reaching the bottom row, and the propagation delay from the rightmost cell to the leftmost cell in the bottom row takes $4\triangle_{FA}$. Thus, the time needed for summands summation is

$$\triangle_{ss} = 4\triangle_{FA} + 4\triangle_{FA} = 8\triangle_{FA}.$$

In general, to perform an m-by-n unsigned number multiplication we can achieve the time complexity

$$\triangle_{US} = \triangle_{AND} + [(m-1) + (n-1)]\triangle_{FA} \qquad (5.1)$$
$$= \triangle_{AND} + (m+n-2)\triangle_{FA}. \qquad (5.2)$$

Also, we have area complexity

$$A_{US} = m \times n \times A_{AND} + (m-1) \times n \times A_{FA}.$$

If $m = n$,

$$\triangle_{US} = \triangle_{AND} + (2n-2)\triangle_{FA},$$

and

$$A_{US} = n^2 \times A_{AND} + n(n-1)A_{FA}. \qquad (5.3)$$

With $\triangle_{AND} = 2\triangle_g$ and $\triangle_{FA} = 2\triangle_g$, we have

$$\triangle_{US} = 2\triangle_g + (2n-2) \cdot 2\triangle_g = (4n-2)\triangle_g.$$

The bottom row of FAs in Figure 5.3 can be replaced by a one-level Carry Lookahead Adder, so that the ripple-carry can be avoided and the delay time can be independent of the length of multiplicand. In this case,

$$\triangle_{US} = \triangle_{AND} + (n-1)\triangle_{FA} + \triangle_{CLA}$$
$$= 2n\triangle_g + \triangle_{CLA}$$
$$= (2n+7)\triangle_g.$$

The efficiency can be further improved by the pipelining mechanism if we add latch registers between the rows of FAs. More than one multiplication operation can be overlapped. That is, the addition of the partial product of an earlier multiplication can be overlapped with that of a later multiplication. If the carry-lookahead is considered as one step, then the intermission of two latches should be no shorter than the delay time of any step.

As to the area complexity, with $A_{AND} = 2A_g$ and $A_{FA} = 10A_g$, when $m = n$, we have

$$A_{US} = n^2 A_{AND} + (n(n-1))A_{FA}$$
$$= 2n^2 A_g + (n^2 - n) \cdot 10A_g$$
$$= (2n^2 + 10n^2 - 10n)A_g$$
$$= (12n^2 - 10n)A_g.$$

108 PARALLEL MULTIPLICATION

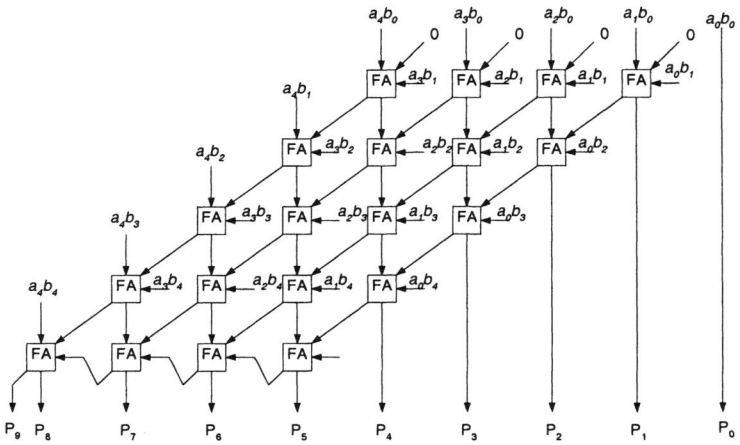

Fig. 5.3: 5×4 Array Multiplier Performing 5-by-5 Multiplication

without CLA. In the case that a CLA is used,

$$\begin{aligned}
A_{US} &= n^2 A_{AND} + (n-1)(n-1) A_{FA} + A_{CLA} \\
&= (2n^2 + 10n^2 - 20n + 10) A_g + A_{CLA} \\
&= (12n^2 - 20n + 10) A_g + A_{CLA}.
\end{aligned}$$

With the LSI or VLSI technology, the above multiplier can be easily implemented on a single chip.

This array is for the unsigned numbers (US) multiplication. For the multipication of sign-magnitude (SM), 1's complement (OC) or 2's complement (TC) numbers, pre-complementer and post-complementer are needed.

Note that A and B are both of $(n+1)$ bits here, and P is of $(2n+1)$ bits, with A_n, B_n and P_{2n} being the sign bits. Recall that a 1's complementer has a delay time of $\triangle_T = 2\triangle_g$ and a 2's complementer has a delay time of $\triangle_T = (2n+2)\triangle_g$. A better solution than the "indirect" array multiplier is sought.

5.3 TWO'S COMPLEMENT ARRAY MULTIPLIER

We introduce in this section a "direct" multiplication method which can multiply 2's complement numbers without pre-complement and post-complement operations. Significant speed up can be provided by a multiplier applying such approach.

We first describe the mathematical principles on which the direct 2's complement multiplication is based. Let $N = (a_{n-1} a_{n-2} \cdots a_0)$ be a 2's complement number where a_{n-1} is the sign bit. The value of N can be expressed as

$$N = \begin{cases} +(a_{n-2}\cdots a_0)_2 = +(\sum_{i=0}^{n-2} a_i 2^i)_{10} & \text{if } a_{n-1} = 0 \\ -[(\bar{a}_{n-2}\cdots \bar{a}_0) + 1]_2 = -(\sum_{i=0}^{n-2}(1-a_i)2^i + 1)_{10} & \text{if } a_{n-1} = 1. \end{cases}$$

That is, for $N = (0a_{n-2}\cdots a_0)_2$, we have

$$N = \left(+\sum_{i=0}^{n-2} a_i 2^i\right) = -a_{n-1} 2^{n-1} + \left(\sum_{i=0}^{n-2} a_i 2^i\right)$$

with the first term equal to 0 since $a_{n-1} = 0$.

For $(N = 1a_{n-2}\cdots a_0)_2$, we have

$$\begin{aligned} N &= -\left(\sum_{i=0}^{n-2}(1-a_i)2^i\right) + 1\right) \\ &= -\left(\sum_{i=0}^{n-2} 2^i - \sum_{i=0}^{n-2} a_i 2^i + 1\right) \\ &= -\left[\left(\sum_{i=0}^{n-2} 2^i\right) + 1\right] + \sum_{i=0}^{n-2} a_i 2^i \\ &= -(2^{n-1}) + \left(\sum_{i=0}^{n-2} a_i 2^i\right) \\ &= -a_{n-1} 2^{n-1} + \left(\sum_{i=0}^{n-2} a_i 2^i\right) \end{aligned}$$

with the first term equal to -2^{n-1} since $a_{n-1} = 1$. Notice that

$$\begin{aligned} \sum_{i=0}^{n-2} 2^i &= \frac{2^0(2^{n-1} - 1)}{2 - 1} \\ &= 2^{n-1} - 1, \end{aligned}$$

and hence

$$\sum_{i=0}^{n-2} 2^i + 1 = 2^{n-1}.$$

By now we have obtained a universal expression for N no matter N is a positive number or a negative number,

$$N = -a_{n-1} 2^{n-1} + \left(\sum_{i=0}^{n-2} a_i 2^i\right). \tag{5.4}$$

110 PARALLEL MULTIPLICATION

Next, we would like to show that $+N$ and $-N$ have the following relationship:

$$N = -a_{n-1}2^{n-1} + \left[\sum_{i=0}^{n-2} a_i 2^i\right] + 0$$

$$\updownarrow \qquad \updownarrow \qquad \updownarrow \qquad \updownarrow \qquad (5.5)$$

$$-N = -(1-a_{n-1})2^{n-1} + \left[\sum_{i=0}^{n-2}(1-a_i)2^i\right] + 1.$$

Proof: Intuitively, from

$$N = -a_{n-1}2^{n-1} + \sum_{i=0}^{n-2} a_i 2^i,$$

we have

$$-N = +a_{n-1}2^{n-1} - \sum_{i=0}^{n-2} a_i 2^i.$$

Recalling that

$$\sum_{i=0}^{n-2} 2^i + 1 = 2^{n-1},$$

we subtract the right-hand side of the above equation from the expression of N and then add the left-hand side of it back. That is,

$$-N = -2^{n-1} + a_{n-1}2^{n-1} + \sum_{i=0}^{n-2} 2^i + 1 - \sum_{i=0}^{n-2} a_i 2^i.$$

It can be rewritten as

$$-N = -(1-a_{n-1})2^{n-1} + \left[\sum_{i=0}^{n-2}(1-a_i)2^i\right] + 1.$$

□

From the above example we can find that negating a number can be realized by negating bits, such as changing a_i to $(1 - a_i)$ and changing 0 to 1, or vice versa. This motivates designers to modify a full adder so that some of the inputs and outputs can carry negated weights. Figure 5.4 shows the construction of different types of adders. A Type I adder has no negative input. A Type II has one negative input, and as its modified version, Type II' has two negative inputs. As the modified version of Type I, Type I' has three negative inputs. Also given are the arithmetic relationships between the inputs and outputs of these adders.

With a certain mix of different types of adders, we can construct some array which performs direct complement multiplication.

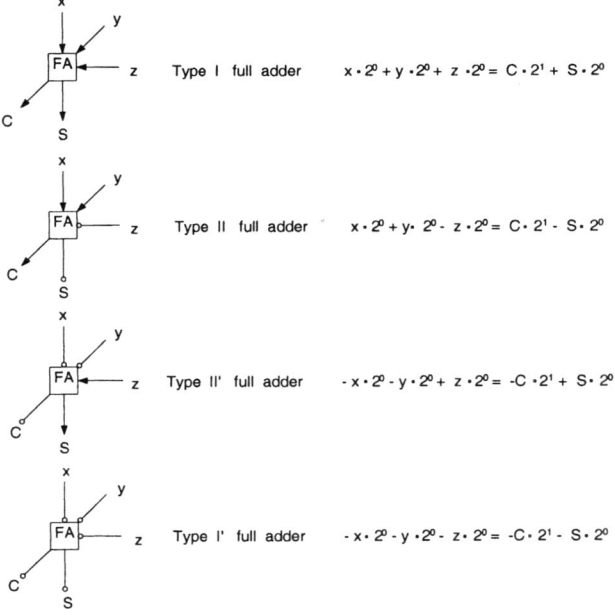

Fig. 5.4: Different Types of Full Adders

5.3.1 Baugh-Wooley Two's Complement Multiplier

Let $A = (a_{m-1}a_{m-2} \cdots a_0)$ be an m-bit 2's complement number, $B = (b_{n-1}b_{n-2} \cdots b_0)$ an n-bit 2's complement number, and $P = A \times B = (p_{m+n-1}p_{m+n-2} \cdots p_0)$ an $(m+n)$-bit product. From Equation (4.1),

$$A = -a_{m-1}2^{m-1} + \left(\sum_{i=0}^{m-2} a_i 2^i\right),$$

$$B = -b_{n-1}2^{n-1} + \left(\sum_{i=0}^{n-2} b_i 2^i\right),$$

and

$$P = -p_{m+n-1}2^{m+n-1} + \left(\sum_{i=0}^{m+n-2} p_i 2^i\right).$$

Since

$$P = A \times B,$$

$$P = \left(-a_{m-1}2^{m-1} + \sum_{i=0}^{m-2} a_i 2^i\right) \times \left(-b_{n-1}2^{n-1} + \sum_{i=0}^{n-2} b_i 2^i\right)$$

$$= + a_{m-1}b_{n-1}2^{m+n-2} + \sum_{i=0}^{m-2}\sum_{j=0}^{n-2} a_i b_j 2^{i+j}$$

$$- \sum_{i=0}^{m-2} a_i b_{n-1} 2^{n-1+i} - \sum_{i=0}^{n-2} a_{m-1} b_i 2^{m-1+i}.$$

Here the third term and the fourth term are negative. We would like the negative weight to be carried by different bits. It can be proved that the third term

$$-\sum_{i=0}^{m-2} a_i b_{n-1} 2^{n-1+i} = 2^{n-1}\left(-2^m + 2^{m-1} + \bar{b}_{n-1}2^{m-1} + b_{n-1} + \sum_{i=0}^{m-2} \bar{a}_i b_{n-1} 2^i\right).$$

Proof: The approach of proof is to show

$$LHS = RHS = \begin{cases} 0 & \text{if } b_{n-1} = 0 \\ -\sum_{i=0}^{m-2} a_i 2^{n-1+i} & \text{if } b_{n-1} = 1, \end{cases}$$

provided that b_{n-1} can only have two values, 0 or 1.
If $b_{n-1} = 0$, we have

$$LHS = 0,$$

$$RHS = 2^{n-1}\left(-2^m + 2^{m-1} + \bar{0} \cdot 2^{m-1} + 0 + \sum_{i=0}^{m-1} \bar{a}_i \cdot 0 \cdot 2^i\right)$$

$$= 2^{n-1}(-2^m + 2^{m-1} + 2^{m-1})$$

$$= 0.$$

If $b_{n-1} = 1$, we have

$$LHS = -\sum_{i=0}^{m-2} a_i 2^{n-1+i},$$

$$RHS = 2^{n-1}\left(-2^m + 2^{m-1} + \bar{1} \cdot 2^{m-1} + 1 + \sum_{i=0}^{m-1} \bar{a}_i \cdot 1 \cdot 2^i\right).$$

$$= 2^{n-1}\left(-1 \cdot 2^m + 1 \cdot 2^{m-1} + 1 + \sum_{i=0}^{m-2} \bar{a}_i 2^i\right)$$

$$\updownarrow \quad \updownarrow \quad \updownarrow \quad \updownarrow \quad \updownarrow \text{ according to Equation (5.5)}$$

$$= -2^{n-1}\left(-0 \cdot 2^m + 0 \cdot 2^{m-1} + 0 + \sum_{i=0}^{m-2} a_i 2^i\right)$$

$$= -2^{n-1} \sum_{i=0}^{m-2} a_i 2^i$$

$$= -\sum_{i=0}^{m-2} a_i 2^{n-1+i}.$$

Hence $LHS = RHS$ for all the cases. □

By rearranging the third term, the RHS of Equation (5.3.1) in the decreasing order of weight, we have

$$-\sum_{i=0}^{m-2} a_i b_{n-1} 2^{n-1+i} = (-1) \cdot 2^{m+n-1} + (1 + \bar{b}_{n-1}) 2^{m+n-2} + \bar{a}_{m-2} b_{n-1} 2^{m+n-3}$$
$$+ \bar{a}_{m-3} b_{n-1} 2^{m+n-4} + \cdots + \bar{a}_0 b_{n-1} 2^{n-1} + b_{n-1} 2^{n-1}.$$

The entries listed below show the values in different weight positions.

2^{m+n-1}	2^{m+n-2}	2^{m+n-3}	2^{m+n-4}	2^{m+n-5}	$\cdots$	2^{n-1}
0	$\bar{b}_{n-1}$	$\bar{a}_{m-2} b_{n-1}$	$\bar{a}_{m-3} b_{n-1}$	$\bar{a}_{m-4} b_{n-1}$	$\cdots$	$\bar{a}_0 b_{n-1}$
(-1)	1	0	0	0	$\cdots$	b_{n-1}

Likewise, the fourth term in P can be written as

$$-\sum_{i=0}^{n-2} a_{m-1} b_i 2^{m-1+i} = 2^{m-1}\left(-2^n + 2^{n-1} + \bar{a}_{m-1} 2^{n-1} + a_{m-1} + \sum_{i=0}^{n-2} a_{m-1} \bar{b}_i 2^i\right).$$

The entries listed below show the values in different weight positions for the fourth term.

2^{m+n-1}	2^{m+n-2}	2^{m+n-3}	2^{m+n-4}	2^{m+n-5}	$\cdots$	2^{m-1}
0	$\bar{a}_{m-1}$	$a_{m-1} \bar{b}_{n-2}$	$a_{m-1} \bar{b}_{n-3}$	$a_{m-1} \bar{b}_{n-4}$	$\cdots$	$a_{m-1} \bar{b}_0$
(-1)	1	0	0	0	$\cdots$	a_{m-1}.

Adding the second row entries of term 3 together with those of term 4, we have

	2^{m+n-1}	2^{m+n-2}	2^{m+n-3}	2^{m+n-4}	2^{m+n-5}	$\cdots$	2^{m-1}	2^{n-1}
	(-1)	1	0	0	0	$\cdots$		b_{n-1}
	(-1)	1	0	0	0	$\cdots$	a_{m-1}	
	(-1)	1	0	0	0	0	$\cdots$	a_{m-1} b_{n-1}
discard.	p_{m+n-1}							

Figure 5.5 shows the summand matrix of the four terms, two of which are negative, in an m-by-n multiplication with $m > n$, applying the above direct 2's complement multiplication algorithm.

114   PARALLEL MULTIPLICATION

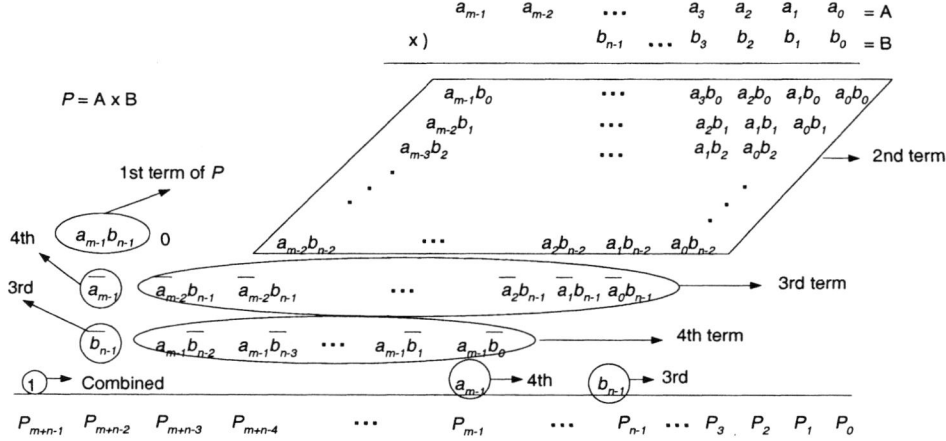

Fig. 5.5: Distribution of Negative Weight

An array implementing such algorithm is referred to as a Baugh-Wooley's array. For an example of $m = 6$ and $n = 4$, a hardware structure is given in Figure 5.6. The structure requires the input bits made available in both complemented form and original form. This is not difficult to satisfy, especially when the bits are held in flip-flops.

The delay time of the array can be calculated by

$$\triangle_{NOT} + \triangle_{AND} + [(n-1) + m]\triangle_{FA}$$
$$= 1 + 2 + (m + n - 1) \times 2\triangle_g$$
$$= [2(m+n) + 1]\triangle_g.$$

Let $A = (10)_{10} = (001010)_2$ and $B = (-3)_{10} = (1101)_2$, we verify the result $P = A \times B = 10 \times (-3) = -30$ by the Baugh-Wooley 2's complement multiplication in Figure 5.7.

For the case of $m = n$, a 4-by-4 Baugh-Wooley 2's complement array multiplier is presented in Figure 5.8. The delay time of such array can be calculated as

$$\triangle_{NOT} + \triangle_{AND} + [(n-1) + (n+1)]\triangle_{FA}$$
$$= (1 + 2 + 2n \times 2)\triangle_g$$
$$= (4n + 3)\triangle_g.$$

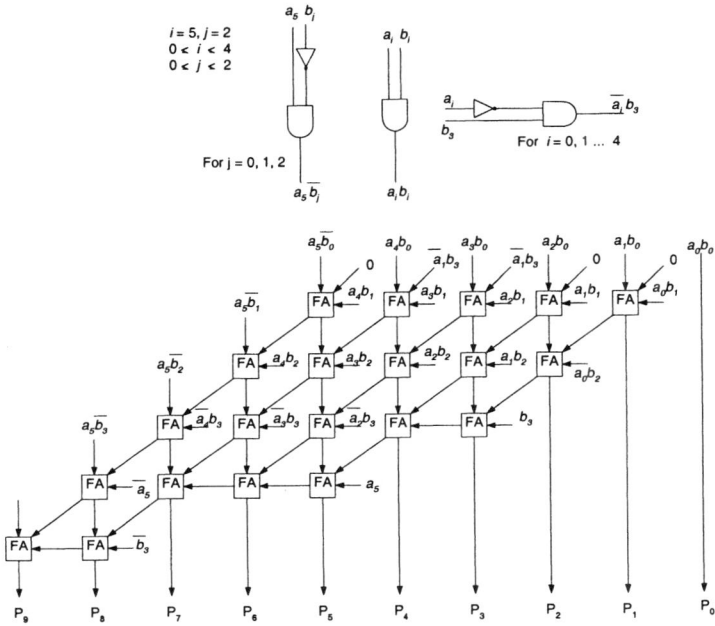

Fig. 5.6: Baugh-Wooley Array Multiplier Performing 6-by-4 Two's Complement Multiplication

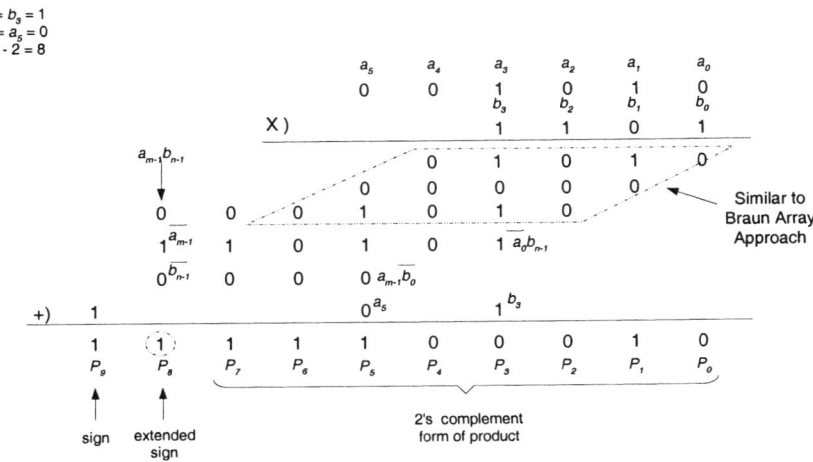

Fig. 5.7: Baugh-Wooley Multiplication for $10 \times (-3)$

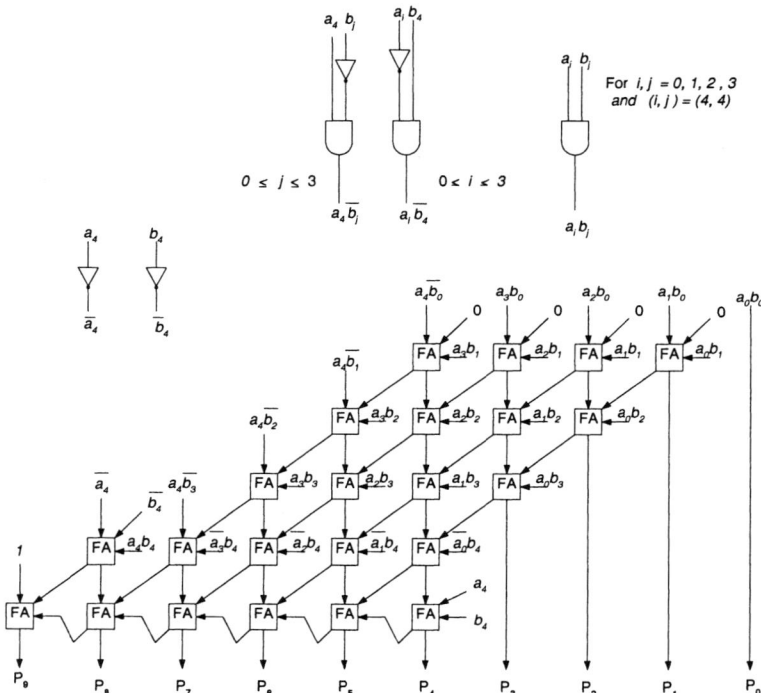

Fig. 5.8: Baugh-Wooley Array with $m=n=5$

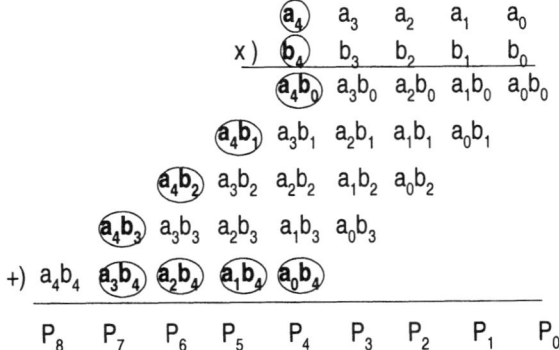

Fig. 5.9: Distribution of the Negative Weight

5.3.2 Pezaris Two's Complement Multipliers

In addition to the Baugh-Wooley 2's complement array multiplier, we describe below another type of direct 2's complement array multiplier, the Pezaris array multiplier and the modified versions of it.

Let $A = \mathbf{a_4}a_3a_2a_1a_0$ be the multiplicand and $B = \mathbf{b_4}b_3b_2b_1b_0$ the multiplier. In the above notations and what follows, a negatively weighted bit is in bold. To multiply A and B, the summands distribution is presented in Figure 5.9.

To implement the above operation, a *Pezaris array multiplier* is proposed, which is composed of Type I, Type I', Type II and Type II' full adders. For a 5-by-5 Pezaris array, the schematic logic circuit diagram is shown in Figure 5.10.

In Figure 5.10 the rightmost Type II adder in the second row from bottom is a little special. As an output of the whole array, the "s" signal produced by that adder is supposed to carry a positive weight. However, in a typical Type II adder,

$$x + y - z = 2c - s$$

is implemented, that is, s carries a negative weight. Since $2c - s = 2c + s - 2s = 2(c - s) + s$, we can let the rightmost Type II adder output an s carrying the positive weight, but subtract s at the input of the rightmost Type II' adder which is one bit position higher, hence it carries twice the weight. Among the x, y and z three inputs of the Type II' adder, z is a negative weight carried input, so z is tied to the s output of the Type II adder on its right. When that output is 0, nothing will affect the Type II' adder on the left.

$n(n-1)$ full adders in total are required in an n-by-n Pezaris array. In general, the delay time for an m-by-n Pezaris array is

$$\begin{aligned} &= \triangle_{AND} + [(n-1) + (m-1)]\triangle_{FA} \\ &= 2\triangle_g + (m+n-2)2\triangle_g \\ &= [2(m+n) - 2]\triangle_g. \end{aligned}$$

118 PARALLEL MULTIPLICATION

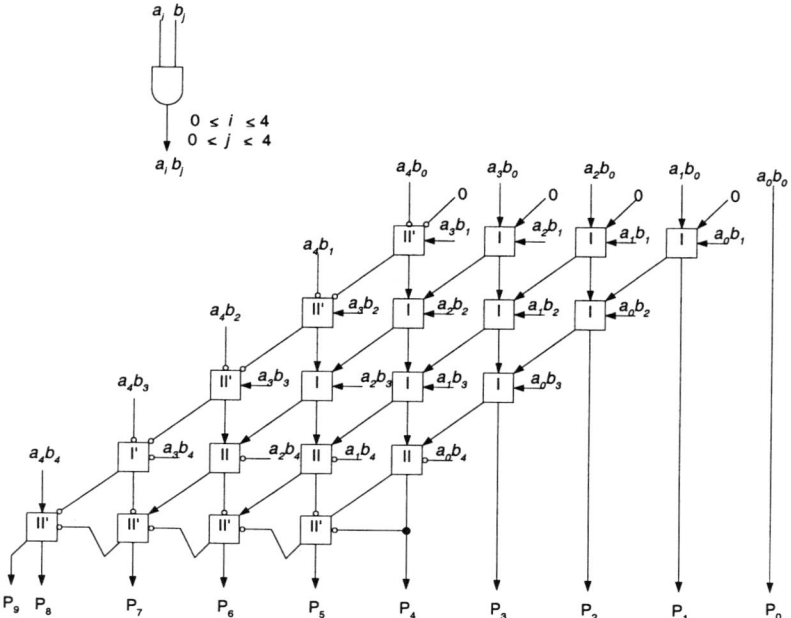

Fig. 5.10: 5-by-5 Pezaris Array Multiplier

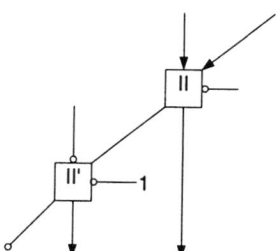

Fig. 5.11: The Adjustment

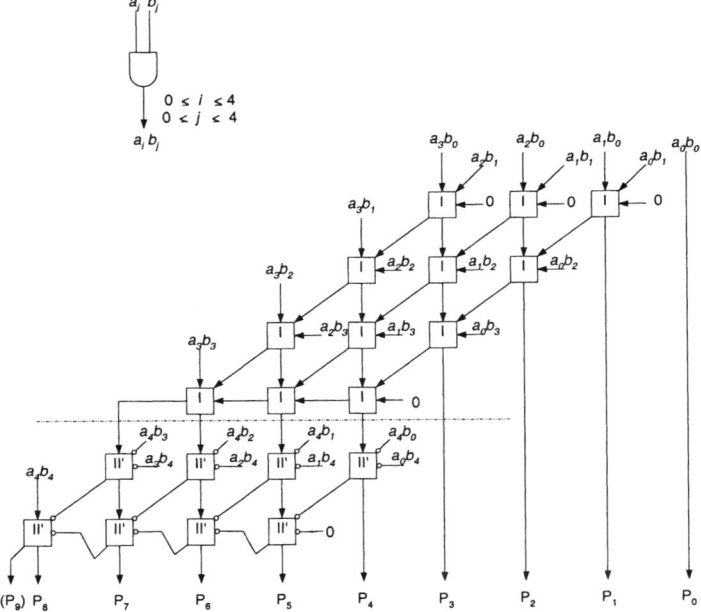

Fig. 5.12: 5-by-5 Bi-Section Array Multiplier

An array multiplier containing fewer types of full adders is proposed in Figure 5.12. It contains only two types of full adders, and is referred to as *bi-section array multiplier*. The Type I full adders are located in the upper section, and Type II' full adders are in the lower section. A 5-by-5 multiplication can be performed by the given bi-section array multiplier.

The delay time for an m-by-n bi-section array multiplier is counted as

$$[n + (m-1)]\triangle_{FA} + \triangle_{AND}$$
$$= [2(m+n-1) + 2]\triangle_g$$
$$= 2(m+n)\triangle_g.$$

A *tri-section array multiplier* consists of three sections. One contains Type I full adders only, one contains solely Type II adders and the whole third section contains Type II' full adders. In Figure 5.13 depicting a 5-by-5 tri-section array, the three sections are separated by dash lines.

In a more general case, for an m-by-n tri-section array multiplier the delay time is

$$[(n-1) + (m-1)]\triangle_{FA} + \triangle_{AND}$$
$$= (m+n-2)2\triangle_g + 2\triangle_g$$
$$= [2(m+n) - 2]\triangle_g.$$

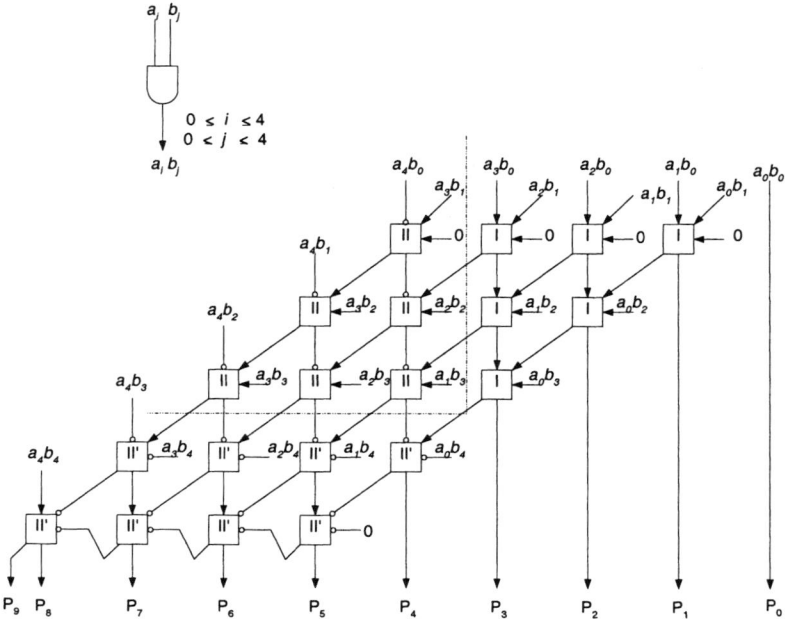

Fig. 5.13: 5-by-5 Tri-section Array Multiplier

5.4 MODULAR STRUCTURE OF LARGE MULTIPLIER

To multiply two numbers of large sizes, a modular structure is needed which can generate and sum sub-products and can have a recursive organization.

5.4.1 Modular Structure

Multiply Modules are array multipliers which can perform fast multiplication on short or moderate-length operands. Given n-by-n multiply modules, we introduce below how to build a large $2n$-by-$2n$ multiplier.

Let $\circ$ denote the function of concatenation. For large multiplicand A, we can partition it into higher order half, A_H, and lower order half, A_L, such that $A = A_H \circ A_L$. Similarly, multiplier B can be partitioned into B_H and B_L with $B = B_H \circ B_L$. If A contains $2n$ bits, A_H and A_L each contains n bits. The same case holds for B. Therefore the $2n$-by-$2n$ multiplication $A \times B$ can be completed by a number of n-by-n multiplication, and each generates a sub-product of $2n$ bits long. That is,

$$P = A \times B \tag{5.6}$$
$$= A_H \circ A_L \times B_H \circ B_L \tag{5.7}$$
$$= (A_H \times B_H + A_L \times B_H) + (A_H \times B_L + A_L \times B_L). \tag{5.8}$$

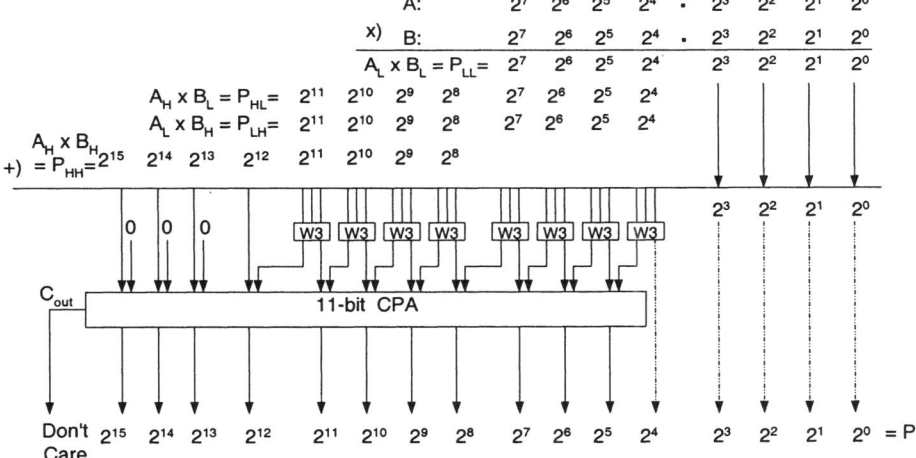

Fig. 5.14: Alignment of the Sub-Products

Fig. 5.15: 8-by-8 Multiplication via 4-by-4 Multipliers

When adding the four sub-products in Equation (5.8), special attention should be paid to the alignment. Note that A_H is n bit position higher than A_L, so the first sub-product $A_H \times B_H$ is n bit position higher than the second sub-product $A_L \times B_H$, and the third sub-product $A_H \times B_L$ is n bit position higher than the fourth sub-product $A_L \times B_L$. In the meantime, since B_H is n bit position higher than B_L, the second sub-product $A_L \times B_H$ is n bit position higher than the fourth sub-product $A_L \times B_L$, hence is in the same position as the third sub-product $A_H \times B_L$.

Figure 5.14 shows the alignment of the four sub-products. Each sub-product can be generated by an n-by-n multiplier module, and four such modules are needed in total. For example, if A and B are 8-bit numbers, the bit position span for the four sub-products is shown in Figure 5.15 indicated by their weights.

To add the sub-products, 3-input Wallace trees are needed for bit position n to $3n - 1$.

122   PARALLEL MULTIPLICATION

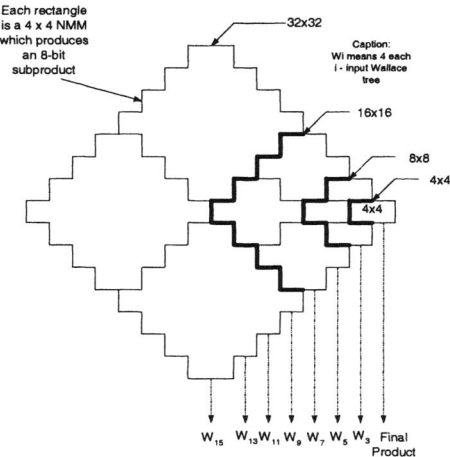

Fig. 5.16: Modular Structure of Array Multipliers

Swapping the order of the first and second sub-products, as well as the third and the fourth, we can have the outputs of the modules arranged in a left-right-top-bottom fashion, as shown in Figure 5.16, where each rectangle is a 4-by-4 module to obtain an 8-by-8 multiplier. Furthermore, such structure can be recursively applied. We can build a 16-by-16 multiplier by putting the 8-by-8 multipliers in left-right-top-bottom way and build a 32-by-32 multiplier in a similar way as shown in Figure 5.16. Different Wallace trees are required depending on the size of the multiplier. We just marked the types of Wallace trees needed in the right half of the multiplier. The left half is symmetric to the right.

The time complexity of such multiply module (MM) can be found as follows based on Equation (5.1) and with an example of 8-by-8 multiplication.

$$\triangle_{MM4 \times 4} = \triangle_{AND} + [(4-1) + (4-1)]\triangle_{FA}$$
$$= (2 + 6 \times 2)\triangle_g$$
$$= 14\triangle_g,$$

$$\triangle_{W_3} = \triangle_{FA} = 2\triangle_g,$$

hence,

$$\triangle_{MM8 \times 8} = (14 + 2)\triangle_g + \triangle_{CPA}$$
$$= 16\triangle_g + \triangle_{CPA}.$$

As to the area complexity, from Equation (5.3) we have

$$A_{MM4 \times 4} = 4^2 A_{AND} + 4 \times 3 A_{FA}$$

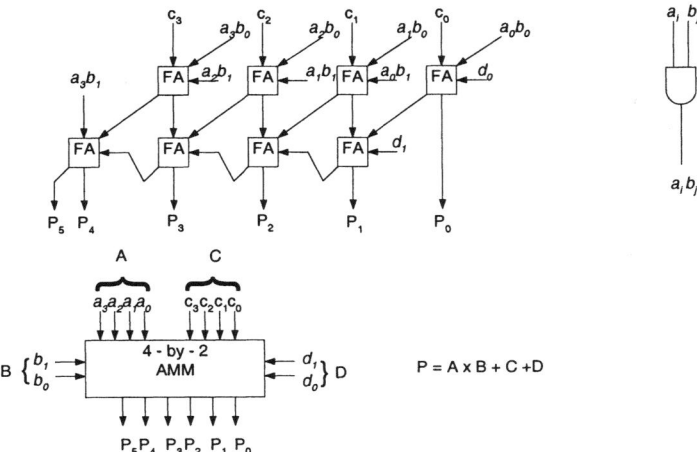

Fig. 5.17: 4-by-2 Additive Multiply Module

$$= 16 \times 2A_g + 12 \times 10A_g$$
$$= 152 A_g,$$

$$A_{W_3} = A_{FA} = 10 A_g,$$

therefore,

$$A_{MM8 \times 8} = (4 \times 152 + 8 \times 10) A_g + A_{CPA}$$
$$= 688 \triangle_g + A_{CPA}.$$

5.4.2 Additive Multiply Modules

The *additive multiply modules* (AMM) can receive additional addends and add them to the product of the input multiplicand and multiplier. The multiply modules mentioned in the prior subsection have no such function and are referred to as *non-additive multiply modules*. Figure 5.17 shows a 4-by-2 AMM which can perform the arithmetic operation such as $P = A \times B + C + D$. Here A and B are a 4-bit multiplicand and 2-bit multiplier respectively. The product P is of 6 bits since $4 + 2 = 6$. C is of 4 bits, and D is of 2 bits.

Let

$$P = A \times B$$
$$= A_H \circ A_L \times B_4 \circ B_3 \circ B_2 \circ B_1.$$

That is,

124 PARALLEL MULTIPLICATION

```
                        A:       2⁷  2⁶  2⁵  2⁴ . 2³  2²  2¹  2⁰
               x)       B:       2⁷  2⁶. 2⁵  2⁴ . 2³  2² . 2¹  2⁰
                                         2⁵  2⁴  2³  2²  2¹  2⁰     =A_L x B₁
                            2⁹  2⁸  2⁷  2⁶  2⁵  2⁴                   =A_H x B₁
                                2⁷  2⁶  2⁵  2⁴  2³  2²               =A_L x B₂
                    2¹¹ 2¹⁰  2⁹  2⁸  2⁷  2⁶                           =A_H x B₂
                            2⁹  2⁸  2⁷  2⁶  2⁵  2⁴                   =A_L x B₃
            2¹³ 2¹²  2¹¹ 2¹⁰  2⁹  2⁸                                  =A_H x B₃
                    2¹¹ 2¹⁰  2⁹  2⁸  2⁷  2⁶                           =A_L x B₄
    +)  2¹⁵ 2¹⁴  2¹³ 2¹²  2¹¹ 2¹⁰                                     =A_H x B₄

        2¹⁵ 2¹⁴  2¹³ 2¹²  2¹¹ 2¹⁰  2⁹  2⁸  2⁷  2⁶  2⁵  2⁴  2³  2²  2¹  2⁰
```

Fig. 5.18: 8-by-8 Multiplication via 4-by-2 Multipliers

$$
\begin{array}{r}
A_H \circ A_L \\
\times) \quad B_4 \circ B_3 \circ B_2 \circ B_1 \\ \hline
A_L \times B_1 \\
A_H \times B_1 \\
A_L \times B_2 \\
A_H \times B_2 \\
A_L \times B_3 \\
A_H \times B_3 \\
A_L \times B_4 \\
+) \quad A_H \times B_4 \\ \hline
P
\end{array}
$$

For example, if A and B are 8-bit numbers, the bit position span for the four subproducts is shown in Figure 5.18, indicated by their weights.

The array multiplier implemented by 4-by-2 AMMs is presented in Figure 5.19. As we can see from Figure 5.17, the worst case delay for an 4-by-2 AMM is

$$
\begin{aligned}
\triangle_{AMM(4 \times 2)} &= 5\triangle_{FA} + \triangle_{AND} \\
&= (10+2)\triangle_g \\
&= 12\triangle_g.
\end{aligned}
$$

Notice that if only P_1 and P_0 are needed, it is not necessary to wait for $12\triangle_g$ nothing rather, $6\triangle_g$ is sufficiently long. Hence in Figure 5.19, the pair of AMMs connected by the dashed lines encounter a delay time of $(6+12=18)\triangle_g$. There are three such

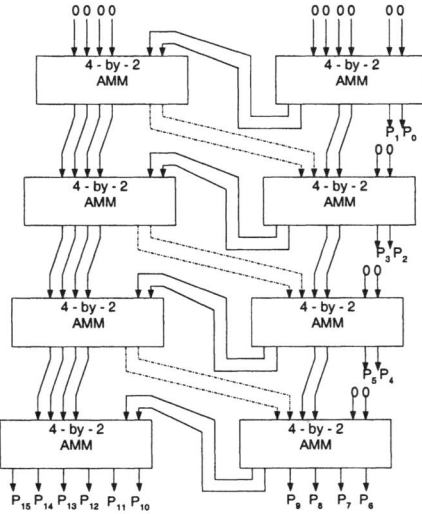

Fig. 5.19: Modular Structure Applying Additive Multiply Modules

pairs along the delay path. Including the first and last AMMs, the total delay time of the modular multiplier completing 8-by-8 multiplication is

$$\triangle_{AMM(8\times 8)} = 12 + 18 + 18 + 18 + 12 = 78\triangle_g.$$

The area complexity of this 8-by-8 modular array can be calculated as follows:

$$\begin{aligned} A_{AMM(4\times 2)} &= 8A_{FA} + 8A_{AND} \\ &= (80 + 16)A_g \\ &= 96A_g. \end{aligned}$$

In total,

$$A_{AMM(8\times 8)} = 8 \times A_{AMM(4\times 2)} = 8 \times 96A_g = 768A_g.$$

5.4.3 Programmable Multiply Modules

Observe the summand matrix in Figure 5.20 (a), and partition it into two halves horizontally and vertically. We have the sub-matrixes in four regions indicated as (0), (1), (2) and (3). One can find that the sub-matrix in region (0) contains no negatively weighted summand. The sub-matrix in (1) has the summands in the bottom row carrying negative weight, and that in (3) has the summands on the left border carrying negative weight. In region (2), the sub-matrix has the features of both matrixes in (1) and (3). Let AMM_i denote the type of Additive Multiply Module needed in region i.

126   PARALLEL MULTIPLICATION

Figure 5.20 (a) implies that a 2's complement multiplier can be composed of four additive multiply modules: AMM_0, AMM_1, AMM_2 and AMM_3. Furthermore, with these kind of AMMs, even a bigger multiplier can be built. See Figure 5.20 (b).

Actually, the array structure of AMM_i can be fixed for different i. Only the input summands to the bottom and/or to the left border need to be altered to make different AMM_is. Rather than inputing $a_i b_j$ all the time, $\bar{a}_i b_j$ is needed at the bottom for AMM_1, and $a_i \bar{b}_j$ is needed on the left border for AMM_3. As to AMM_2, both the bottom and the left border take the input summands specified as above respectively.

In other words, if we can let the summands generation be controlled by function modes, different AMM_is can be formed by switching the mode. When all the summands are generated under mode 0, the array formed is AMM_0. When the summands to the bottom are generated under mode 1, but others are under mode 0, the array formed is AMM_1. Also, when the summands to the left border are generated under mode 3, but others are under mode 0, the array formed is AMM_3. If the summands to the left are generated under mode 3 and that to the bottom under mode 1, and the rest are under mode 0, AMM_2 is formed. In this way the additive multiply modules are made *programmable*.

The summands input to the bottom and left border need one variable to be negated under some mode, otherwise they should be $a_i b_j$ as usual. The mode controlled generation of these summands is shown in Figure 5.21. Note that the FA at the intersection of bottom row and left border accepts both of the summands sent to the former and the latter. Refer to the FA on its left as the "*corner*" based on the Baugh-Wooley's multiplication. The summands input to it under different modes are listed, as well as the additional summands input only during Mode 2 including $\bar{a}_{n-1}$, $\bar{b}_{n-1}$, and a "1".

Figure 5.22 shows how the AMM(4 × 4)s are applied to construct larger multiplication networks, AMM(8×8). Note that $AMM_0(4\times 4)$, $AMM_1(4\times 4)$, $AMM_2(4\times 4)$ or $M_3(4 \times 4)$ are obtained by programming the $AMM_i(4\times 4)s$. This iterative method can be extended to design $(16 \times 16), (32 \times 32), \cdots, (4k \times 4k)$ multipliers with k being an integer. In general, a $(4k \times 4k)$ multiplier requires k^2 $AMM(4\times 4)$s, among which $(k-1)^2$ are $AMM_0(4 \times 4)$, $(k-1)$ are $AMM_1(4 \times 4)$, $(k-1)$ are $AMM_3(4 \times 4)$, and one is $AMM_2(4 \times 4)$.

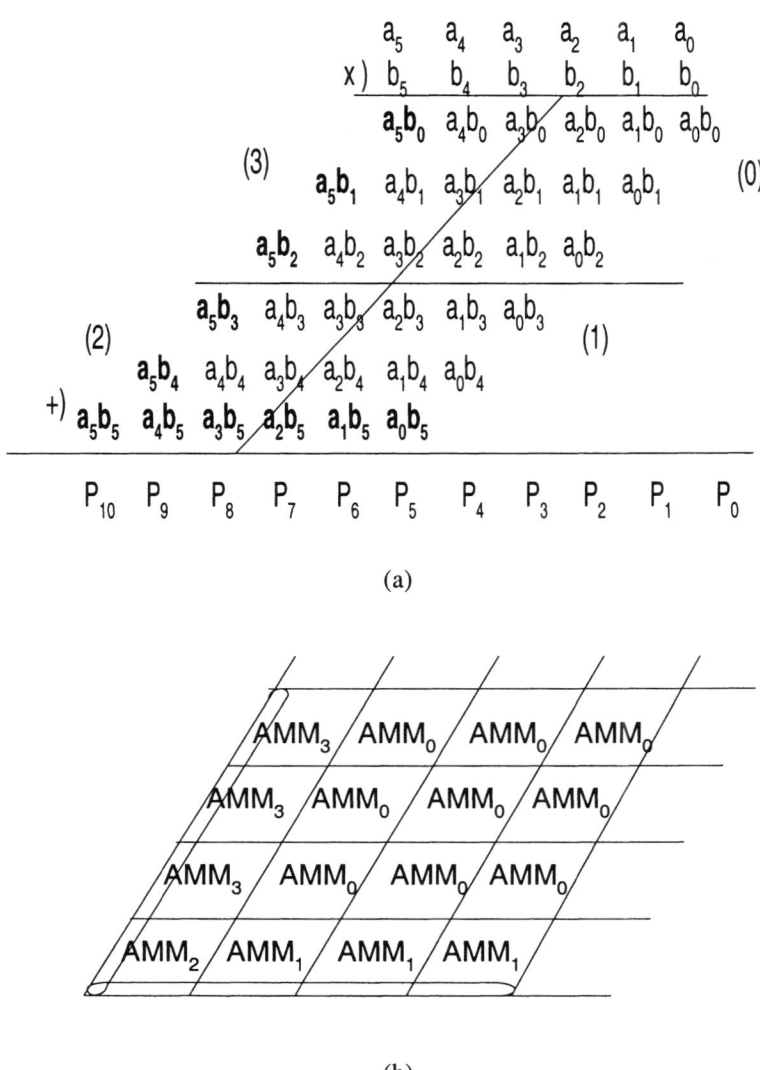

Fig. 5.20: Combine Small AMMs into a Large One

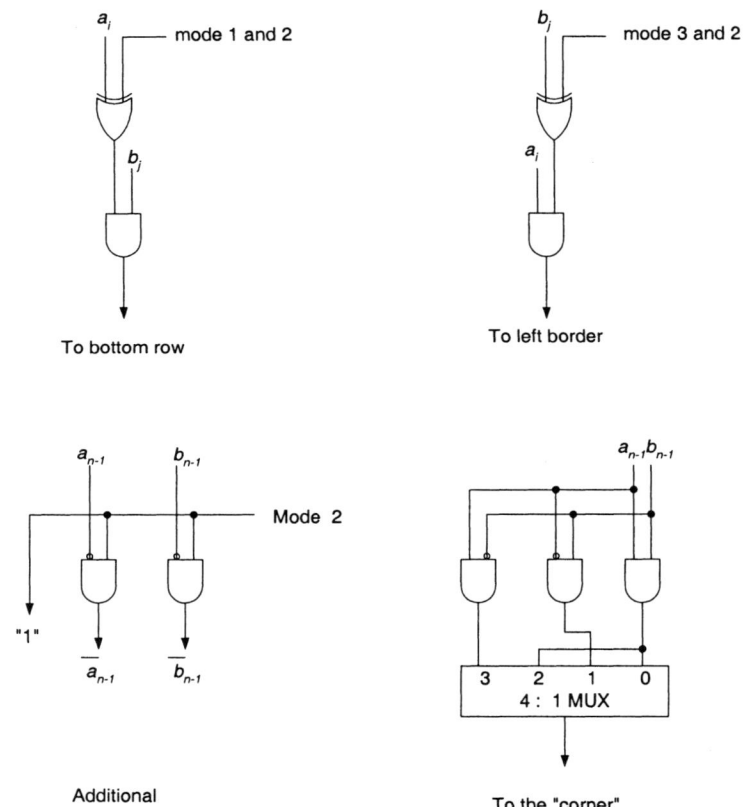

Fig. 5.21: Summands of Preparation in Programmable AMM

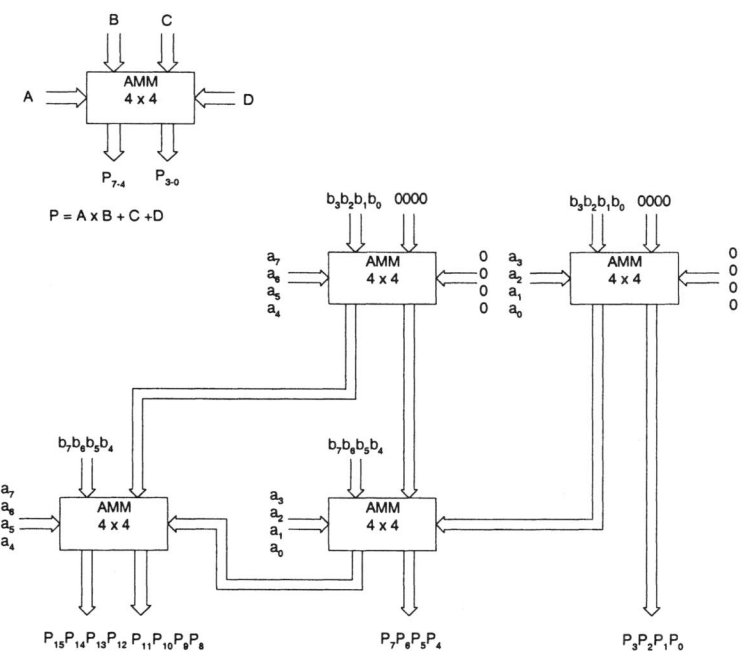

Fig. 5.22: AMM 8 × 8 Applying AMM 4 × 4

REFERENCES

1. Advanced Micro Devices, "TTL/MSI AM2505 4-bit by 2-bit 2's Complement Multiplier," 901 Thompson Place, Sunnyvale, CA.

2. D. P. Agrawal, "Optimum Array-Like Structures for High-Speed Arithmetic," *Proc. 3rd Symposium on Computer Arithmetic*, IEEE Computer Society, #75C1017, Nov. 1975, pp. 208–219.

3. S. F. Anderson, J. G. Earle, R. E. Goldschmidt and D. M. Powers, "The IBM System/360 Model 91: Floating-Point Execution Unit," *IBM J. R & D*, Vol. 11, No. 1, Jan. 1967, pp. 34–53.

4. S. Bandyopadhyay, S. Basu and A. K. Choudhory, "An Iterative Array for Multiplication of Signed Binary Numbers," *IEEE Trans. Comp.*, Vol. C-21, No. 8, Aug. 1972, pp. 921–922.

5. C. R. Baugh and B. A. Wooley, "A Two's Complement Parallel Array Multiplication Algorithm," *IEEE Trans. Comp.*, Vol. C-22, No. 12, Dec. 1973, pp. 1045–1047.

6. A. D. Booth, "A Signed Binary Multiplication Technique," *Quart. J. Mech. Appl. Math.*, Vol. 4, pt. 2, June 1951, pp. 236–240.

7. E. L. Braun, *Digital Computer Design*, Academic Press, New York, 1963.

8. T. A. Brubaker and J. C. Becker, "Multiplication Using Logarithms Implemented with Read-Only Memory," *IEEE Trans. Comp.*, Vol. C-24, 1975.

9. W. Buchholz, ed., *Planning a Computer System*, McGraw-Hill, New York, 1962, pp. 210–211.

10. Y. Chu, *Digital Computer Design Fundamentals*, McGraw-Hill, New York, 1962.

11. L. Dadda, "On Parallel Digital Multipliers," *Alta Frequenza* 45 (1976), pp. 574–580.

12. L. Dadda, "Some Schemes for Parallel Multipliers," *Alta Frequenza*, Vol. 34, Mar. 1965, pp. 349–356.

13. K. J. Dean, "Design of a Full Multiplier," *Proc. IEEE*, Vol. 115, Nov. 1968, pp. 1592–1594.

14. I. D. Deegan, "Cellular Multiplier for Signed Binary Numbers," *Electronic Letters*, Vol. 7, 1971, pp. 436–437.

15. J. Deverrell, "Pipeline Iterative Arithmetic Arrays," *IEEE Trans. Comp.*, C-24 (Mar. 1975), pp. 317–322.

16. Fairchild Semiconductors, "TTl/MSI 9344 Binary (4-bit by 2-bit) Full Multiplier," 313 Fairchild Dr., Mountain View, CA, 1971.

17. I. Flores, *The Logic of Computer Arithmetic*, Prentice-Hall, Englewood Cliffs, NJ, 1963.

18. J. A. Gibson and R. W. Ginbard, "Synthesis and Comparison of Two's Complement Parallel Multipliers," *IEEE Trans. Comp.*, C-24 (1975), pp. 1020–1027.

19. H. H. Guild, "Fully Iterative Fast Array for Binary Multiplication and Fast Addition," *Electronic Letters*, Vol. 5, May 1969, pp. 263.

20. A. Habibi and P. A. Wintz, "Fast Multipliers," *IEEE Trans. Comp.*, C-19 (Feb. 1970), pp. 153–157.

21. Y. Harata, et al. "A High Speed Multiplier Using a Redundant Binary Adder Tree," *IEEE J. Solid-State Circuits* SC-22 (Feb. 1987), pp. 28–33.

22. A. Hemel, "Making Small ROMs Do Math Quickly, Cheaply and Easily," *Electronic Computer Memory Technology*, W. B. Riley, ed., McGraw-Hill, New York, 1971, pp. 133–140.

23. Hughes Aircraft Co., "Bipolar LSI 8-bit Multiplier H1002MC," 500 Superior Avenue, Newport Beach, CA, 1972.

24. K. Hwang, "Global Versus Modular Two's Complement Array Multipliers," *IEEE Trans. Comp.*, Vol. C-28, No. 4, Apr. 1979, pp. 300–306.

25. O. L. MacSorley, "High-Speed Arithmetic in Binary Computers," *Proc. IRE*, Vol. 49, No. 1, Jan. 1961, pp. 67–91.

26. J. C. Majithia and R. Kita, "An Iterative Array for Multiplication of Signed Binary Numbers," *IEEE Trans. Comp.*, C-20 (Feb. 1971), pp. 214–216.

27. G. W. McIver, et al., "A Monolithic 16×16 Digital Multiplier," *Dig. Tech. Paper Int. Solid State Circuits Conf.*, Feb. 1974, pp. 54–55.

28. M. Mehta, V. Parmar, and E. Swartzlander, "High-Speed Multiplier Design Using Multi-input Counter and Compressor Circuits," *Proc. 10th Symp. on Computer Arithmetic* (1991), pp. 43–50.

29. J. N. Mitchell, "Computer Multiplication and Division Using Binary Logarithms," *IRE Trans. Elec. Comp.*, Vol. EC-11, Aug. 1962, pp. 512–517.

30. R. D. Mori, "Suggestion for an IC Fast Parallel Multiplier," *Electronic Letters*, Vol. 5, Feb. 1969, pp. 50–51.

31. F. J. Mowle, *A Systematic Approach to Digital Logic Design*, Addison Wesley, Reading, MA, 1976.

32. T. G. Noll, et al., "A Pipelined 330-MHz Multiplier," *IEEE J. Solid-State Circuits*, 21 (June 1986), pp. 411–416.

33. V. Peng, S. Samudrala and M Gavrielov, "On the Implementation of Shifters, Multipliers and Dividers in VLSI Floating-Point Units," *Proc. 8th Symp. on Computer Arithmetic* (May 1987), pp. 95–102.

34. S. D. Pezaris, "A 40-ns 17-bit by 17-bit Multiplier," *IEEE Trans. Comp.*, Vol. C-20, No. 4, Apr. 1971, pp. 442–447.

35. G. W. Reitwiesner, *Binary Arithmetic in Advances in Computers*, F.L. Alt, ed., Academic, New York, (1960), pp. 231–308.

36. J. E. Robertson, "Two's Complement Multiplication in Binary Parallel Computers," *IRE Trans. Elec. Comp.*, Vol. EC-4, No. 3, Sept. 1955, pp. 118–119.

37. L. P. Rubinfield, "A Proof of the Modified Booth's Algorithm for Multiplication," *IEEE Trans. Comp.*, C-24, (Oct. 1975), pp. 1014–1015.

38. E. A. Swartzlander, Jr., "The Quasi-Serial Multiplier," *IEEE Trans. Comp.*, Vol. C-22, No. 4, Apr. 1973, pp. 317–321.

39. M. R. Sanitoro and M. A. Horowitz, "SPIM, A Pipelined 64×64 Iterative Multiplier," *IEEE J. Solid-State Circuits*, 24 (Apr. 1989), 487–493.

40. TRW, "MPY-LSI Multipliers: AJ 8×8, 12×12 and 16×16," LSI Products, TRW, Redondo Beach, CA, Mar. 1977.

41. C. S. Wallace, "A Suggestion for a Fast Multiplier," *IEEE Trans. Elec. Comp.*, Vol. EC-13, No. 1, Feb. 1964, pp. 14–17.

42. D. Zuras and W. H. McAllister, "Balanced Delay Trees and Combinatorial Division in VLSI," *IEEE J. Solid-State Circuits*, SC-21 (Oct. 1986), 814–819.

PROBLEMS

5.1 (a) Build a 9 bit-slice Wallace tree using carry-save full adders. What is the delay time of this Wallace tree?
(b) Build a 31 bit-slice Wallace tree for which 15-bit Wallace trees can be used in the first level. What is the delay time of this Wallace tree?

5.2 Perform unsigned integers multiplication $A \times B$ using an unsigned array multiplier.
(a) $A = (a_4 a_3 a_2 a_1 a_0)_2$ and $B = (b_6 b_5 b_4 b_3 b_2 b_1 b_0)_2$.
(b) $A = (100101)_2$ and $B = (110011)_2$.

5.3 The Baugh-Wooley algorithm was applied to perform a 6-by-4 two's complement multiplication. After obtaining the following summand matrix, the multiplier was lost due to some fault. Can you (a) Retrieve the multiplier? (b) Complete the multiplication?

			0	0	1	0	1	0
×)				?	?	?	?	
			0	1	0	1	0	
		0	1	0	1	0		
0	0	0	0	0	0	0		
1	1	0	1	0	1			

5.4 Construct a Baugh-Wooley array multiplier to perform 19 by (-7).
(a) Present the summand matrix in the similar way shown in Figure 5.8.
(b) Calculate the time complexity and area complexity of this array multiplier.

5.5 What is the delay time for an m-by-n ($m > n$) Baugh-Wooley array multiplier? Justify your conclusion. Perform 10101×00101 by Pezaris array multiplier. Find the outputs of each cell that lead to the final answer.

5.6 Construct a 32-by-32 bit array multiplier with 4-by-4 non-additive multiply modules.
(a) Show the modular structure of the multiplier.
(b) Find the time complexity and area complexity. Implement a 16-by-16 bit array multiplier through 8-by-4 additive multiply modules (AAMs). Show modular structure in between the AMMs.

5.7 Realize a 16-by-16 AAM via 8-by-8 AMMs.

5.8 Given 2×2 Non-additive Multiply Modules, perform $(1101)_2 \times (0101)_2$. Fill the partial products below. Draw a complete Wallace tree to add the partial products

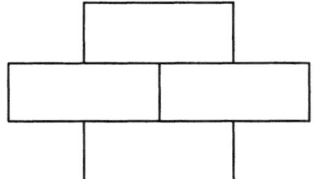

with each FA indicated by the input and output values, and a 5-bit CPA showing the connected inputs and the obtained final result.

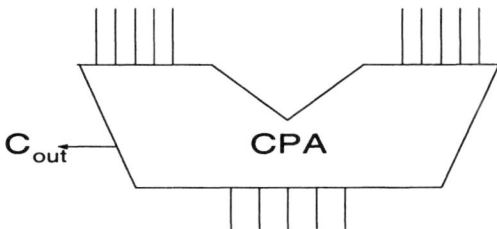

6

Sequential Division

Unlike addition, subtraction and multiplication, division does not, in general, produce an exact answer, because the dividend is not necessarily a multiples of the divisor. A variety of division algorithms are to be described in the rest of this chapter.

6.1 SUBTRACT-AND-SHIFT APPROACH

Let's look at an example of pencil-and-paper division in Figure 6.1.
Here 2746 is referred to as the *Dividend*, 32 is the *Divisor*, 85 the *Quotient* and 26 the *Remainder*. After $274 - 256$ is performed in the first step, 18 is obtained as a *partial remainder*. By pulling down 6, 18 becomes 180 (+6). On digital computers, most of the division operations are performed in a recursive procedure represented by the following formula:

$$R^{(j+1)} = r \times R^{(j)} - q_{j+1} \times D,$$

where $j = 0, 1, \cdots, n-1$ is the *recursion index*, $R^{(j)}$ is the partial remainder in the jth iteration. The initial partial remainder $R^{(0)}$ equals the dividend, and $R^{(n)}$ is the *final remainder*. The quotient is determined digit by digit in the recursive procedure, with q_{j+1} being the $(j+1)$th *quotient digit*. D is the divisor, and r is the radix. In the above example $r = 10$.

From the recursion formula, we have

For $j = 0$: $\quad R^{(1)} = r \times R^{(0)} - q_1 \times D,$
For $j = 1$: $\quad R^{(2)} = r \times R^{(1)} - q_2 \times D$
$\quad\quad\quad\quad\quad = r \times (r \times R^{(0)} - q_1 \times D) - q_2 \times D$

```
            85
       _____
   32 | 2746      2746 ──▶ Dividend
        256        32 ──▶ Divisor
        ___
        186        85 ──▶ Quotient
        160
        ___
         26        26 ──▶ Remainder
```

Fig. 6.1: Pencil-and-Paper Division

$$= r^2 \times R^{(0)} - r \times q_1 \times D - q_2 \times D$$
$$= r^2 \times R^{(0)} - (r \times q_1 + q_2) \times D,$$
For $j = n - 1$: $R^{(n)} = r \times R^{(n-1)} - q_n \times D$
$$\vdots$$
$$= r^n \times R^{(0)} - (r^{n-1} \times q_1 + \cdots + r \times q_{n-1} + q_n) \times D.$$

Let's look at how the quotient digit q_j is determined step by step in the above pencil-and-paper procedure. In the first step, we made an estimation about how many times of 32 is close to but less than 274. We figured out that 8 is probably the right number, since 9 times 32 is over 280 and 274 is not big enough to subtract it. Unlike a human, a machine cannot make such estimation by a first look, but can only perform what it is programmed to do. We can let it repeatedly subtract 32 from 274 until the difference left is no greater than 32. The number of successful subtractions can be recorded, which gives the quotient digit q_j. After shifting the partial remainder, we can begin the trial process for the next quotient digit. Here subtract and shift are most of the operations required, and the division can be completed by the subtract-and-shift approach. In the above example, 85 is not the final result of 2746/32. Instead,

$$\frac{2746}{32} = 85 + \frac{26}{32}.$$

In general,

$$\frac{R^{(0)}}{D} = Q + \frac{R^{(n)}}{D},$$

where

$$Q = \sum_{j=1}^{n} r^j \times q_j.$$

Conventionally, the dividend is $2n$ bits long, and the divisor is of n bits. The word length of the quotient is also n, which tells us when to stop the recursive procedure.

In the long division form shown in Figure 6.2, which represents the conventional case, both the dividend $R^{(0)}$ and the divisor D are assumed to be fractions. That is,

$$R^{(0)} = r_0.r_1r_2\cdots,$$

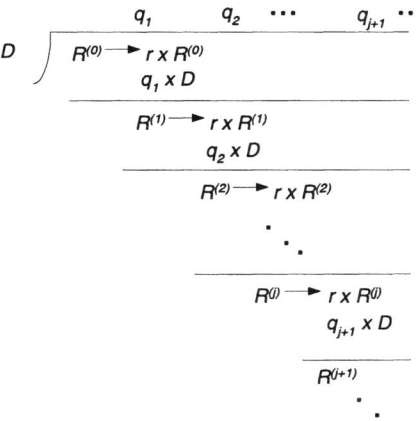

Fig. 6.2: Long Division Form

$$D = d_0.d_1d_2\cdots,$$

where the points on the right of r_0 and d_0 are radix points, and r_0 and d_0 are the signs. Also the quotient Q is a fraction such that

$$Q = q_0.q_1q_2\cdots q_{n-1}q_n,$$

with q_0 being the sign equal to $r_0 \oplus d_0$. It is assumed that $R^{(0)} < D$ and $D \neq 0$, and both are positive numbers for illustration. Here Q is supposedly < 1, and quotient overflow occurs if $R^{(0)} \geq D$.

The operations in Figure 6.3 is performed as follows:

$$R^{(1)} = r \times R^{(0)} - q_1 \times D$$
$$0.48 = 6.08 - 5.60$$

$$R^{(2)} = r \times R^{(1)} - q_2 \times D$$
$$0.6 = 4.8 - 4.2.$$

Actually, in the binary number system, 0 and 1 are the only possible digits. That greatly simplifies the quotient digit selection process.

We will see later that for the selection of the quotients $q_j, j = 1, \cdots, n$, which are based on $R^{(0)}, R^{(1)}, \cdots, R^{(j+1)} \cdots R^{(n-1)}$, different arithmetic conditions will give different procedures. Letting $R^{(j+1)}$ be the partial remainder, we have

- Restoring division : $0 \leq R^{(j+1)} < D$

- Nonrestoring division : $\mid R^{(j+1)} \mid \leq \mid D \mid$

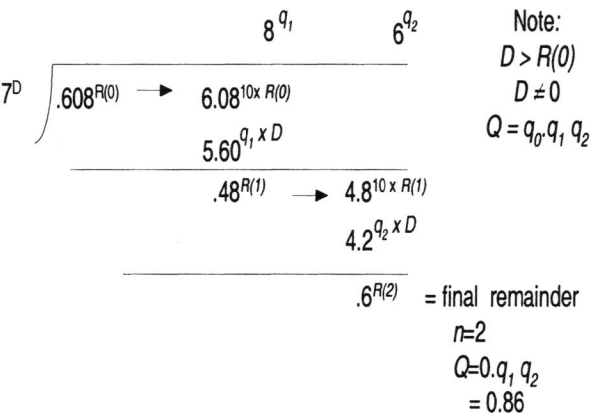

Fig. 6.3: Example of Long Division

- Generalized SRT division : $\mid R^{(j+1)} \mid \leq k \times \mid D \mid$, $1/2 \leq k \leq 1$.

6.2 BINARY RESTORING DIVISION

In the conventional restoring division, $0 \leq R^{(j+1)} < D$. The quotient digit q_{j+1}, $j = 0, 1 \cdots n - 1$, is selected by performing a sequence of subtractions and shifts. Each time D is subtracted form the partial remainder $r \times R^{(j)}$, until the difference becomes negative. Then D is added back to that negative difference, which is so called *restoring*. The last subtraction is canceled by the addition here. Then we have the quotient digit, which is determined by the number of subtractions as $q_{j+1} =$ (number of subtractions -1). In general: $q_{j+1} + 1$ subtractions and 1 addition are required to find q_{j+1} in the worst case. Figure 6.4 shows an example. For a binary number system in which $r = 2$, the worst case scenario can be greatly improved.

$$R^{(j+1)} = r \times R^{(j)} - q_{j+1} \times D$$

becomes

$$R^{(j+1)} = 2R^{(j)} - q_{j+1} \times D,$$

with $q_{j+1} \in \{0,1\}$, $0 \leq R^{(j+1)} < D$. The quotient digit can be determined as follows:

$$q_{j+1} = \begin{cases} 0 & \text{if } 2R^{(j)} < D \\ 1 & \text{if } 2R^{(j)} \geq D. \end{cases}$$

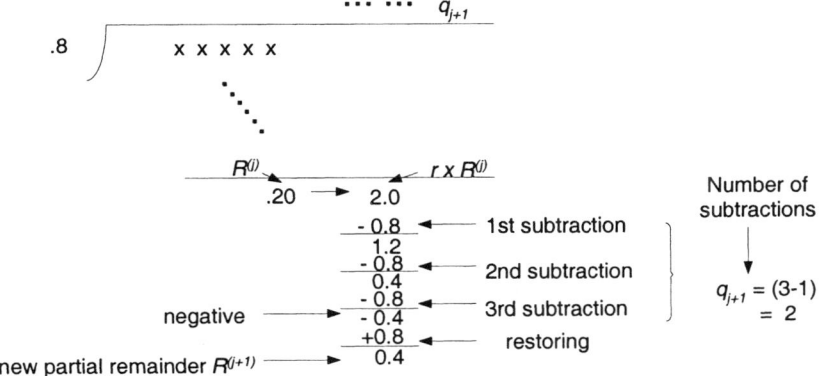

Fig. 6.4: Example of Restoring Procedure

The partial remainder can be obtained by one left shift of $R^{(j)}$, and the trial process can be implemented by one subtraction:

$$R^{(j+1)} = 2R^{(j)} - D.$$

Then we check the "sign" of $R^{(j+1)}$. If it is positive, $q_{j+1} = 1$, else $q_{j+1} = 0$ and one restoring addition is to be performed. So the binary restoring division requires at most one subtraction and one addition to determine one quotient digit. The addition is needed to restore the correct partial remainder:

$$R^{(j+1)} = R^{(j+1)} + D = 2R^{(j)}.$$

The hardware implementation of this binary restoring division is shown in Figure 6.5.

Three registers are included each with n bits: (1) *accumulator* (AC), (2) *auxiliary register* (AX) and (3) the *quotient-multiplier register* (QM), which was used to store the multiplier in multiplication as mentioned in Chapter 4, and is to hold the final quotient in division operations under discussion. The $2n$-bit dividend is initially stored in registers AC concatenating QM. The n-bit divisor is stored in the AX register.

The content of AC concatenating QM can be left shifted. The quotient digit q_{j+1} can be shifted in from the right end, and the bit shifted out from the left end of AC can be stored in a *buffer* register T.

We can see that the similar hardware performing multiplication can be applied for division except the differences made on the following issues. First, the direction of register shift was shift to right in multiplication, while in division it is shift to left. Second, the adder actually performs subtraction now, that is add the 2's complement number of the divisor. In Figure 6.5, the input $\overline{AX}$ to the right side of the Adder is

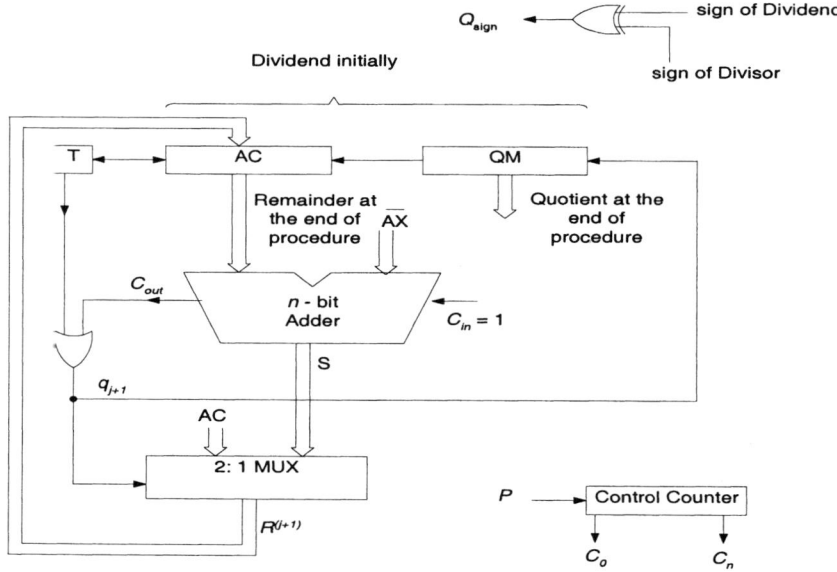

Fig. 6.5: Hardware for Restoring Division

the 1's complement of the dividend. With the C_{in} set to 1, we have $AC + \overline{AX} + 1 = AC - AX$ to inspect whether $2R^{(j)}$ is $< D$ or $\geq D$. C_{out} is the carry out. Quotient digit q_{j+1} is obtained by

$$q_{j+1} = T \vee C_{out},$$

which can be explained as follows. Let $n = 4$, for example.
When $T = 1$, with "x" being an arbitrary digit, we have

$$\begin{array}{rl} 1\ \text{xxxx} & \leftarrow 2R^{(j)} \\ -)\ \ 1010 & \leftarrow D,\ \text{for example} \\ \hline \cdot\ \cdot\ \cdot\ \cdot\ \cdot\ \cdot & \end{array}$$

Obviously, $2R^{(j)} \geq D$ and we should set $q_{j+1} = 1$.
When $T = 0$, we have

$$\begin{array}{rl} 0\ \text{xxxx} & \leftarrow 2R^{(j)} \\ -)\ \ 1010 & \leftarrow D,\ \text{for example} \\ \hline C_{out}\ \cdot\ \cdot & \end{array}$$

If $C_{out} = 1$, that means $2R^{(j)} + (2^n - D) \geq 2^n$, which is equivalent to $2R^{(j)} \geq D$. Hence we should set $q_{j+1} = 1$. So, $q_{j+1} = 1$ if $T = 1$ "or" $C_{out} = 1$.

When $D < 2R^{(j)}$, a restoring operation occurs. The subtraction is canceled by addition, and the partial remainder is kept unchanged. In this case $q_{j+1} = 0$. Instead of subtracting D and then adding it back, a 2-to-1 MUX is used which can select the unchanged partial remainder to route through when the control signal $q_{j+1} = 0$. The other input data of the MUX is the difference of $2R^{(j)} - D$, which will be selected when $q_{j+1} = 1$.

As the partial remainder shifts out bit by bit from the left of register $AC \circ QM$, more and more bit positions are available on the right for the quotient. The space for partial remainder is reduced from the original $2n$ bits to n bits, and that for the quotient grows to n bits. The final remainder $R^{(n)}$ will be in $T \circ AC_1 AC_2 \cdots AC_{n-1}$ standing for the value $R^{(n)} \times 2^{-n}$.

The compare-and-shift operation should be repeated n times, and loading the input operands needs one cycle. So the binary restoring division requires $n + 1$ cycles in total.

6.3 BINARY NON-RESTORING DIVISION

An improved division method is the binary non-restoring division which does not need the "restoring addition" mentioned previously. The assumption that $D > 0$ and $|R^{(j+1)}| < D$ remain the same, while the partial remainder, $R^{(j+1)}$, is allowed to have either a positive or a negative value. The operation to be performed can be either subtraction or addition, depending on the partial remainder. That is,

$$R^{(j+1)} = \begin{cases} 2R^{(j)} - D & \text{if } 2R^{(j)} > 0 \\ 2R^{(j)} + D & \text{if } 2R^{(j)} < 0. \end{cases}$$

Accordingly, the quotient digit can be either $+1$ or -1 but not 0, that is, $q_{i+1} \in \{1, -1\}$. The quotient digit selection is described as follows:

$$q_{(j+1)} = \begin{cases} 1 & \text{if } 0 < 2R^{(j)} < D \\ -1 & \text{if } -D < 2R^{(j)} < 0. \end{cases}$$

When $2R^{(j)} = 0$, the process can be terminated. As a result the quotient Q is represented by signed-digit code that contains no zeros. -1 is denoted as $\bar{1}$.

In the restoring division, when $2R^{(j)} - D < 0$ the remainder is restored to $2R^{(j)}$. After shifting it and subtracting D again, $4R^{(j)} - D$ is obtained. In the non-restoring division, when $2R^{(j)} - D < 0$ we stay with the negative difference and correct it by adding D in the next iteration. That is $2(2R^{(j)} - D) + D = 4R^{(j)} - D$, the same result as above can be obtained. If two consecutive quotient digits $q_j q_{j+1}$ are selected as 01 in the restoring division, they appear as $1\bar{1}$ in the non-restoring division. Obviously $01 = 1\bar{1}$.

The binary non-restoring division (B.N.D.) of $0.001011 \div 0.01$ is performed in Figure 6.6(a) in contrast to the binary restoring division (B.R.D.) of that shown in (b).

In the binary non-restoring division, no restoration is needed and thus the MUX in Figure 6.5 can be eliminated. Consequently, the division time is improved. Noticing that

$$\triangle_T(B.N.D) \quad \alpha \quad n \times (\text{subtraction or addition})$$
$$\triangle_T(B.R.D) \quad \alpha \quad n \times (\text{subtraction} + \text{restoration}),$$

we have

$$\triangle_T(B.R.D.) > \triangle_T(B.N.D.).$$

The quotient obtained in the signed digit representation may need to be converted to a form compatible with other operands such as 2's complement form, which can be easily implemented as follows. Let the sum of weights of all the positive digits be Q^+, and that of all the negative digits be Q^-, we have

$$Q^+ + Q^- = Q \qquad (6.1)$$
$$Q^+ - Q^- = 1 - 2^{-n} \qquad (6.2)$$

since for whichever position containing a 0 in Q^+, there exists a $\bar{1}$ in Q^-, hence $Q^+ - Q^- = 0.11 \cdots 1$. Adding both sides of Equation (6.1) and Equation (6.2), we have

$$2Q^+ = Q + 1 - 2^{-n}$$

or

$$Q = 2Q^+ - 1 + 2^{-n}.$$

$2Q^+$ can be obtained by left shifting Q^+ for one bit position, -1 means change the sign to its opposite after the shift, and $+2^{-n}$ means insert a 1 in the LSB position which is vacated after the shift. Q^- is ignored here.

In the middle of the division, if a partial remainder turns to be 0, it means that the subsequent quotient digits are all 0s. A 0 remainder detection circuitry may be designed so that the algorithm can halt immediately if a 0 remainder is encountered.

Conventionally the final remainder is of the same sign as the dividend. If $7 \div 4$ ends with $Q = 2$ and $R = -1$, which is correct, an adjustment should be made by $Q = 2 - 1 = 1$ and $R = -1 + 4 = 3$. That is, $Q = Q - 2^{-n}$ and $R = R + D$ if the dividend and the divisor have the same sign in the binary non-restoring division. If the dividend and divisor have the opposite signs, then the adjustment can be done by $Q = Q + 2^{-n}$ and $R = R - D$.

Both the restoring division and non-restoring division involve a large number of shifting which is proportional to n, as well as r. To further speed up the division time high-radix division can be adopted. We will discuss the technique in this regard in the next section.

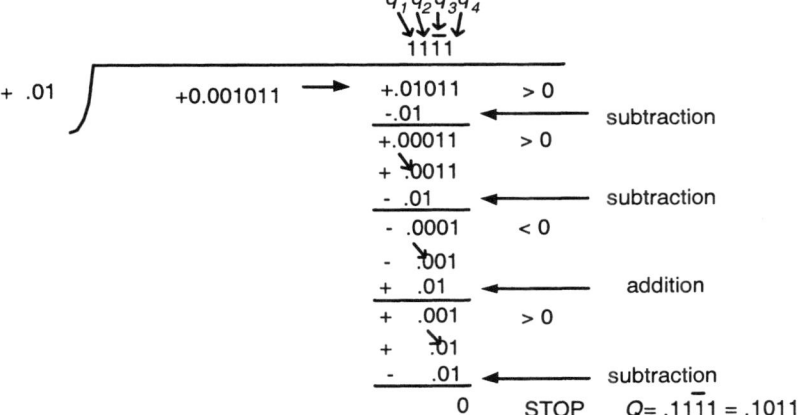

(a) Binary Non-Restoring Division

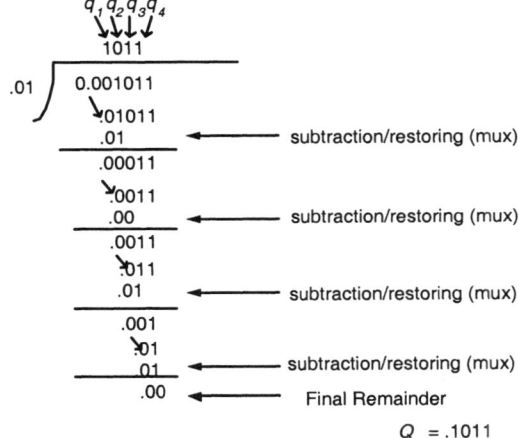

(b) Binary Restoring Division

Fig. 6.6: Division Performed by Non-Restoring/Restoring Algorithms

6.4 HIGH-RADIX DIVISION

Previously described division methods are radix 2 divisions, in which radix $r = 2$, and digit set $\in \{0, 1\}$, which is formed by 1-bit strings. In radix 4 division $r = 4$, and digit set $\in \{00, 01, 10, 11\}$, which is formed by 2-bit strings. In radix 8 division $r = 8$, and digit set $\in \{000, 001, 010, 011, 100, 101, 110, 111\}$, which is formed by 3-bit strings, and so forth. For $r \geq 2$, the case is referred to as high-radix division.

As mentioned before, high-radix division yields faster execution time, since the number of iterations needed for a fixed sized dividend is reduced. However, we will learn in the following subsections that this enhanced speed is at the expense of increased hardware.

6.4.1 High-Radix Non-Restoring Division

Recall that the quotient digit set for binary non-restoring division is $\{-1, 1\}$, but not 0. For the high-radix non-restoring division, the quotient digit set is $\{-(r-1), \cdots, -1, 1, \cdots, (r-1)\}$, where r is the radix.

For example, $r = 4$, radix-4 non-restoring division quotient digit set

$$\in \{-3, -2, -1, 1, 2, 3\}.$$

Again, here,

$$-D < R^j < D,$$

that is, $-4D < 4R^{(j)} < 4D$ as before. The multiple choices for q_{j+1} are:

$$q_{j+1} = \begin{cases} -3 & \text{if } -4D < 4R^{(j)} < -3D \\ -3 \text{ or } -2 & \text{if } -3D < 4R^{(j)} < -2D \\ -2 \text{ or } -1 & \text{if } -2D < 4R^{(j)} < -D \\ -1 & \text{if } -D < 4R^{(j)} < 0 \\ +1 & \text{if } 0 < 4R^{(j)} < D \\ +1 \text{ or } +2 & \text{if } D < 4R^{(j)} < 2D \\ +2 \text{ or } +3 & \text{if } 2D < 4R^{(j)} < 3D \\ +3 & \text{if } 3D < 4R^{(j)} < 4D. \end{cases}$$

Note that the quotient found by the non-restoring division is a radix-r signed-digit code, that is,

$$Q = .q_1 q_2 \cdots q_n$$

with each $q_i \in \{-(r-1), \cdots, -1, 1, \cdots, (r-1)\}$.

Therefore we may need to change a radix-r signed-digit number back to a conventional radix-complement form such as

$$Q = q_0.q_1' q_2' \cdots q_n'$$

with each $q'_i \in \{0, 1, \cdots, (r-1)\}$ and

$$q_0 = \begin{cases} 0 & \text{for a positive number} \\ 1 & \text{for a negative number.} \end{cases}$$

The following algorithm shows how to convert a signed-digit number into a radix-complement number.

1. $Q = q_1 q_2 \cdots q_n$;

2. if $q_1 < 0$, {
   $q'_1 \leftarrow r + q_1$;
   $q_0 \leftarrow r - 1$;
   }

3. else {
   $q'_1 \leftarrow q_1$;
   $q_0 \leftarrow 0$;
   }

4. $j \leftarrow 1$;

5. if $q_{j+1} < 0$, {
   $q'_j \leftarrow q_j - 1$;
   $q'_{j+1} \leftarrow r + q_{j+1}$;
   }

6. else {
   $q'_j \leftarrow q_j$;
   $q'_{j+1} \leftarrow q_{j+1}$;
   }

7. if $j \neq n - 1$, {
   $j \leftarrow j + 1$;
   go to 5;
   }

8. else $Q' \leftarrow (q_0.q'_1 q'_2 \cdots q'_n)$;
   Stop.

A serial inspection of the quotient digits is required in this conversion procedure. A positive digit q_{j+1} should remain unchanged, a negative digit should increase by r and borrow 1 from the higher order digit.

6.4.2 SRT Division

SRT is named after its inventors Sweeney, Robertson and Tocher. Independently and at about the same time, D. W. Sweeney of IBM, J. E. Robertson of the University of Illinois and K. D. Tocher of Imperial College, London, discovered a new method of binary division. In their method,

$$\frac{1}{2} < |D| < 1$$

is assumed, which means that the divisor is a normalized fraction in the form $0.1d_2 \cdots d_n$. Also assumed is

$$\frac{1}{2} < |2R^{(j)}| < 1,$$

which means that all the partial dividends are normalized fractions. Recall that the partial dividend is r times of the partial remainder in general. Unlike in non-restoring division algorithm $q_{j+1} \in \{-1, 1\}$,

$$q_{j+1} \in \{-1, 0, 1\}$$

now.

Here the divisor is shifted, or added to or subtracted from the partial dividend since $\{-1, 0, 1\}$ is to be selected. That is,

$$R^{j+1} = \begin{cases} 2R^{(j)} + D, & \text{if } 2R^{(j)} < -D, \\ 2R^{(j)}, & \text{if } -D \leq 2R^{(j)} \leq D, \\ 2R^{(j)} - D, & \text{if } 2R^{(j)} > D. \end{cases}$$

The rule for selecting the quotient digit q_{j+1} is as follows:

$$q_{j+1} = \begin{cases} -1 & \text{if } 2R^{(j)} < -D, \\ 0 & \text{if } -D \leq 2R^{(j)} \leq D, \\ 1 & \text{if } < 2R^{(j)} > D. \end{cases}$$

Notice that both the divisor, D, and the partial dividend, $2R^{(j)}$, are normalized fractions, or $\frac{1}{2} < |D| < 1$ and $\frac{1}{2} < |2R^{(j)}| < 1$. By using $\frac{1}{2} < |D|$, the comparison is reduced to

$$R^{j+1} = \begin{cases} 2R^{(j)} + D, & \text{if } 2R^{(j)} \leq -\frac{1}{2}, \\ 2R^{(j)}, & \text{if } -\frac{1}{2} < 2R^{(j)} \leq \frac{1}{2}, \\ 2R^{(j)} - D, & \text{if } 2R^{(j)} \geq \frac{1}{2}, \end{cases}$$

and the quotient selection rule can be accordingly simplified as

$$q_{j+1} = \begin{cases} -1, & \text{if } |2R^{(j)}| > \frac{1}{2}, \text{ and sign of } 2R^{(j)} \text{ is negative}, \\ 0, & \text{if } |2R^{(j)}| \leq \frac{1}{2}, \\ +1, & \text{if } |2R^{(j)}| > \frac{1}{2}, \text{ and sign of } 2R^{(j)} \text{ is positive}. \end{cases}$$

The advantage of using this new set of rules to determine $R^{(j+1)}$ and q_{j+1} lies in the fact that only comparisons of $2R^{(j)}$ against the constant $\frac{1}{2}$ or $-\frac{1}{2}$ are required.

6.4.3 Modified SRT Division

Wilson and Ledley proposed a modified version of the SRT division to further reduce the number of addition and subtraction operations in divisions. This method assumes that the divisor D is positive and a normalized fraction, and the dividend N is either normalized or with at most a single zero after the binary point. $N < D$ holds as before. The algorithm is described by the flow chart in Figure 6.7.

The iterative procedure is entered by

$$N^{(s)} = N^{(s-1)} - D.$$

In one iteration, α quotient digits can be determined if in the partial remainder $N^{(s)}$ there are α 0s to the right of the binary point.

If $N^{(s)} < 0$, which will be the case for the first iteration due to the above assumption, set the new quotient digit q_i to 0, and attach $(\alpha - 1)$ 1s to its right. The next iteration process will perform $N^{(s)} = N^{(s-1)} + D$.

If $N^{(s)} \geq 0$, set q_i to 1 and attach $(\alpha - 1)$ 0s to its right. The next iteration process will perform $N^{(s)} = N^{(s-1)} - D$.

The quotient digit index is updated by $i = i + \alpha$ in each iteration, and when it reaches the desired number of bits in the quotient the process ends.

Figure 6.8 shows a numerical example for verifying the described procedure. The quotient obtained by this method will be in minimal but not necessarily canonical form given normalized divisors.

6.4.4 Robertson's High-Radix Division

Let's generalize the SRT division which was originally proposed by Robertson, with an arbitrary radix r, and a quotient digit set

$$q_{j+1} \in \{-m, \cdots, -1, 0, 1, \cdots, m\}$$

where $\frac{1}{2}(r-1) \leq m \leq r-1$. The successive quotient digits selections should satisfy the recursive formula

$$R^{(j+1)} = r \times R^{(j)} - q_{j+1} \times D,$$

such that after the subtraction, the partial remainder always has

$$|R^{(j+1)}| \leq k \times |D|, \tag{6.3}$$

with k being a constant no greater than 1. Division methods exist for certain discrete value of k in the range

$$\frac{1}{2} \leq k \leq 1. \tag{6.4}$$

When $k = 1$, it becomes the non-restoring division introduced in section 6.2. A straightforward but lengthy analysis of all the cases that may arise reveals that

$$q_{j+1} \leq k(r-1). \tag{6.5}$$

148 SEQUENTIAL DIVISION

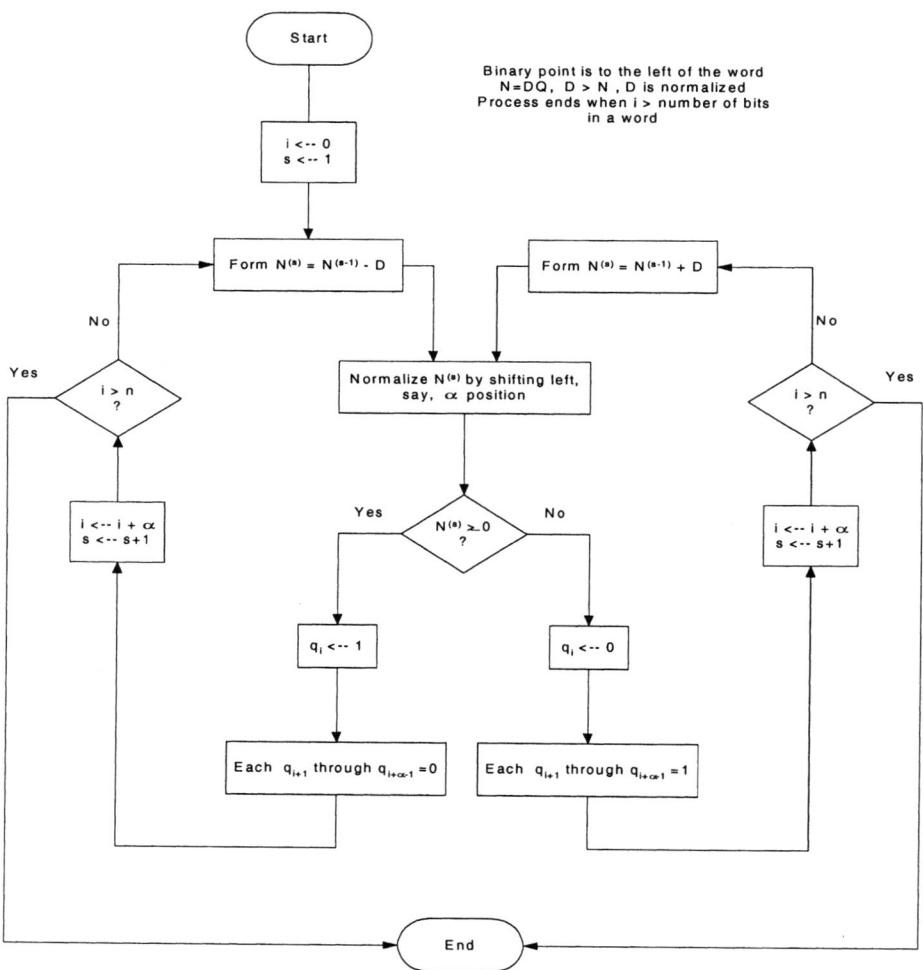

Fig. 6.7: Flow Chart for Wilson-Ledley's Division Algorithm

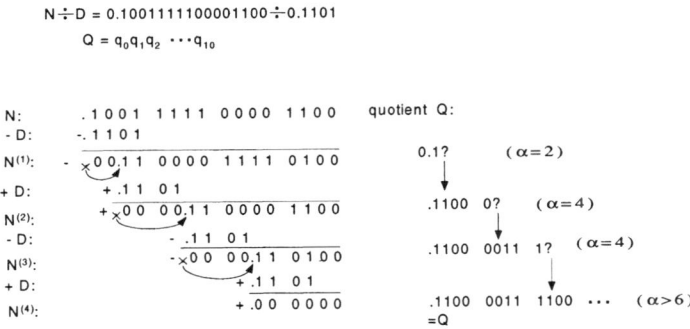

Fig. 6.8: Numerical Example for Wilson-Ledley's Division Algorithm

The Robertson's division method includes the following three steps:

Step 1. The partial remainder $R^{(j)}$ is left shifted for one digital position.
Step 2. One of several permissible arithmetic procedures is selected, such that the maximum absolute value of $r|R^{(j)}|$ is reduced by the amount of $1/r$. Note that due to inequality (6.3),

$$r|R^{(j)}| \leq rk|D|.$$

For the same reason, the next partial remainder $R^{(j+1)}$ satisfies

$$|R^{(j+1)}| \leq k \times |D|.$$

Step 3. A quotient digit is generated corresponding to the arithmetic procedure selected.

The arithmetic procedure is the key to realize the reduction of $\frac{1}{r}$ mentioned in Step 2. That is, to select a proper q_{j+1} in $R^{(j+1)} = r \times R^{(j)} - q_{j+1} \times D$ to transform $r \cdot R^{(j)}$ into $R^{(j+1)}$. Let

$$q_{j+1} = -m, \cdots, -2, -1, 0, +1, +2, \cdots, m$$

and

$$\lfloor \frac{r-1}{2} \rfloor < m < (r-1).$$

Plot in Cartesian system $R^{(j+1)}$ versus $rR^{(j)}$. When $q_{j+1} = 0$, $R^{(j+1)} = rR^{(j)}$ which is corresponding to a line of 45 degree slope through the origin. For different discrete values of q, a family of lines with such slope, referred to as q-lines, are in correspondence. From inequality (6.3), R^{j+1} should be bounded by $k|D|$, and $rR^{(j)}$ bounded by $rk|D|$. Recall that $\frac{1}{2} \leq k \leq 1$ in Equation (6.4). Hence a rectangle with vertices $(rk|D|, k|D|)$, $(rk|D|, -k|D|)$, $(-rk|D|, -k|D|)$ and $(-rk|D|, k|D|)$ can be formed and the q-lines are confined within the rectangle. Figure 6.9(a) shows the above specified diagram which is referred to as Robertson's diagram.

The number of q-lines should be minimized so that fewer multiples of the divisor need to be prepared. On the other side, the overlap of projections of the q-lines should be maximized to assure the precision. The two requirements are contradictory. Considering that the rightmost q-line should go through the upper right vertex of the rectangle, we plug $(rk|D|, k|D|)$ into

$$R^{(j+1)} = r \times R^{(j)} - m \times |D|$$

to solve for k, and find that

$$k = \frac{m}{r-1}.$$

Since $k \geq \frac{1}{2}$ (from (Equation 6.4)), we have

$$m \geq \frac{r-1}{2}.$$

On the other hand, $q_j \not> r - 1$, and as a possible value of q_j, $m \not> r - 1$. m is thereby restricted to the range

$$\frac{r-1}{2} \leq m \leq r - 1.$$

The binary non-restoring division is a special case of Robertson's high-radix division, and the Robertson's diagram for it is shown in Figure 6.9(b). Since $k = 1$, the boundary of the rectangle becomes D and rD. Since $m = 1$, and $q_{j+1} = 0$ is not allowed, there are only two q-lines in the diagram.

The Robertson's diagram for the binary restoring division is shown in Figure 6.9(c). Since $q_{j+1} \in \{0, 1\}$, two q-lines exist and one of them goes to the original point. Since only positive remainders are allowed, the two q-lines do not extend below the horizontal axis.

6.5 CONVERGENCE DIVISION

The division algorithms described in the preceding sections are based on the successive subtraction and shift technique. In this section a different approach is introduced. Iterative multiplications are to be performed to generate the desired quotient. Therefore, the same hardware used for multiplication can be used for division.

In the new proposed method, the dividend and divisor are treated as the numerator and denominator of a fraction. They are multiplied by the same sequence of convergence factors while the value of the fraction remains unchanged. When the denominator approaches 1, the numerator becomes the desired quotient.

CONVERGENCE DIVISION

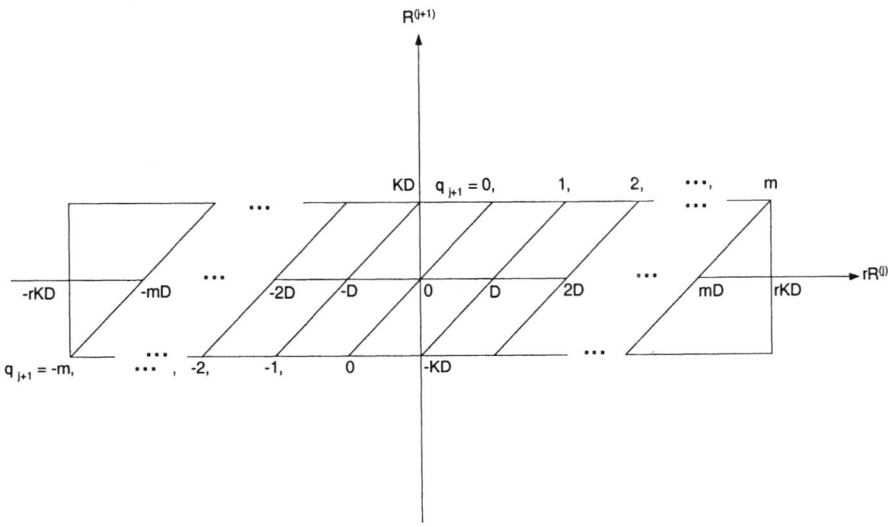

(a) High-Radix Division with $q_{j+1} \in \{-m, \cdots, -2, -1, 0, 1, 2, \cdots, m\}$

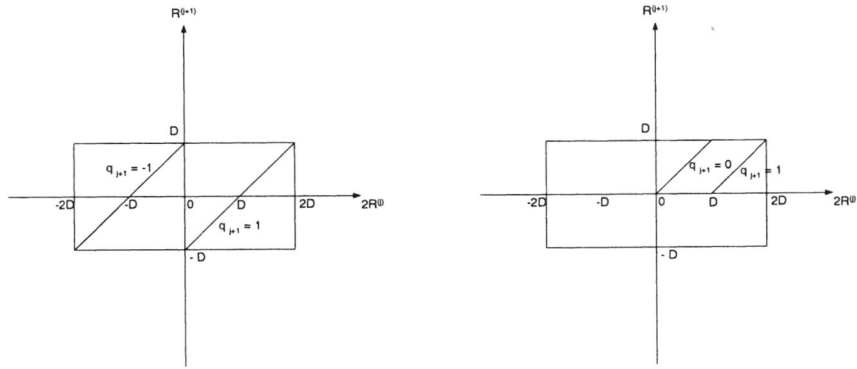

(b) Binary Non-Restoring Division with $q_{j+1} \in \{-1, 1\}$

(c) Binary Restoring Division with $q_{j+1} \in \{0, 1\}$

Fig. 6.9: Robertson Diagrams

6.5.1 Convergence Division Methodologies

Let the division operation be $\frac{N}{D} = Q$, where Q is the quotient or fraction, N is the numerator or dividend and D is the denominator or divisor.

The convergence division approach is to iteratively multiply D and N by a constant factor, P_i for $i = 0, 1, \cdots m$, such that the resulting denominator, $D \cdot (\prod_{i=0}^{i=m} R_i)$ converges to 1 in a quadratic rate (see below) and hence the resulted numerator, $N \cdot (\prod_{i=0}^{i=m} R_i)$ converges toward the desired quotient Q in the quadratic rate. That is,

$$\frac{N}{D} = Q$$

$$\frac{N \times R_0 \times R_1 \times \cdots \times R_m}{D \times R_0 \times R_1 \times \cdots R_m} = Q.$$

For a sufficiently large m,

$$D \times R_0 \times R_1 \times \cdots \times R_m \longrightarrow 1$$
$$N \times R_0 \times R_1 \times \cdots \times R_m \longrightarrow Q.$$

Here the operations carried on is free of division.

The assumption here is that N and D are positive and normalized binary fractions the same as in binary SRT division. That is,

$$\frac{1}{2} \leq N < 1$$
$$\frac{1}{2} \leq D < 1.$$

The general rule to choose the successive multipliers R_i for $i = 0, 1, \cdots, m$ is that

$$D < D_0 < D_1 < D_2 \cdots < D_i$$

and

$$\begin{cases} D_0 = D \times R_0 \\ D_1 = D_0 \times R_1 = D \times R_0 \times R_1 \\ D_2 = D_1 \times R_2 = D \times R_0 \times R_1 \times R_2 \\ \vdots \\ D_i = D_{i-1} \times R_i = D \times R_0 \times \cdots \cdots \times R_i. \end{cases}$$

Finally, when $i = m$,

$$D_i \to 1.0,$$

that is,

$$D_m = D \times R_0 \times \cdots \times R_m \to 1.0.$$

While there could be many ways to choose the factor R_is, we are interested in those of quadratic order or higher, noticing that $(1 + x^n)(1 - x^n) = 1 - x^{2n}$. As assumed, $\frac{1}{2} \leq D < 1$ (D is a normalized fraction), there exists $D + \varepsilon = 1$. We can see that

$$\varepsilon > 0 \quad \text{since} \quad D < 1$$

and

$$\varepsilon \leq \frac{1}{2} \quad \text{since} \quad D \geq \frac{1}{2},$$

so

$$0 < \varepsilon \leq \frac{1}{2}.$$

Note that we have

$$D = 1 - \varepsilon$$

or

$$\varepsilon = 1 - D.$$

Now let's choose the first multiplying factor

$$R_0 = 1 + \varepsilon.$$

Then,

$$\begin{aligned}
D_0 &= D \times R_0 \\
&= D \times (1 + \varepsilon) \\
&= (1 - \varepsilon) \times (1 + \varepsilon) \\
&= 1 - \varepsilon^2 (= 1 - \varepsilon^{2^1}).
\end{aligned}$$

Obviously,

$$D = 1 - \varepsilon < D_0 = 1 - \varepsilon^2.$$

For $i = 1$,

$$D_1 = D_0 \times R_1 \quad (D_i = D_{i-1} \times R_i).$$

Let's choose

$$R_1 = 1 + \varepsilon^2,$$

$$D_1 = D_0 \times (1 + \varepsilon^2)$$
$$= (1 - \varepsilon^2)(1 + \varepsilon^2)$$
$$= 1 - \varepsilon^4 (= 1 - \varepsilon^{2^2}).$$

Obviously,
$$D_0 = 1 - \varepsilon^2 < D_1 = 1 - \varepsilon^4.$$

For general i
$$D_i = D_{i-1} \times R_i.$$

In general, choose
$$R_i = 1 + \varepsilon^{2^i},$$
$$D_i = (1 - \varepsilon^{2^i}) \times (1 + \varepsilon^{2^i})$$
$$= 1 - \varepsilon^{2^{(i+1)}}.$$

Once again,
$$D_{i-1} = 1 - \varepsilon^{2^i} < D_i = 1 - \varepsilon^{2^{(i+1)}}$$

or
$$1 - D_{i-1} = \varepsilon^{2^i} > 1 - D_i = \varepsilon^{2^{(i+1)}}.$$

Note that since
$$0 < \varepsilon \leq \frac{1}{2},$$
$$\varepsilon^{2^i} \leq \left(\frac{1}{2}\right)^{2^i}$$

or
$$\varepsilon^{2^{(i+1)}} \leq \left(\frac{1}{2}\right)^{2^{(i+1)}}.$$

Thus
$$1 - D_i = \varepsilon^{2^{(i+1)}} \leq \left(\frac{1}{2}\right)^{2^{(i+1)}}.$$

As we noticed, $(\frac{1}{2})^{2^{(i+1)}}$ will be a very small number if i is "large". For example, if $i = 5$,
$$\left(\frac{1}{2}\right)^{2^6} = \left(\frac{1}{2}\right)^{64} \cong 5.42 \times 10^{-20};$$

if $i > 9$,
$$\left(\frac{1}{2}\right)^{2^{(i+1)}} \approx 0.$$

That is $\varepsilon^{2^{(i+1)}} = 1 - D_i$, defined as "error", is small and negligible if i is large.

As a result, for large m, $D_m = 1 - \varepsilon^{2^{m+1}} \approx 1$, (note that $D_m = D \times R_0 \times R_1 \times \cdots \times R_m$) and

$$\frac{N \times R_0 \times \cdots \times R_m}{D \times R_0 \times \cdots \times R_m} \approx N \times R_0 \times \cdots \times R_m = Q.$$

So to find Q we just need to let

$$Q = N \times \underbrace{(1+\varepsilon)}_{R_0} \times \underbrace{(1+\varepsilon^2)}_{R_1} \times \cdots \times \underbrace{(1+\varepsilon^{2^m})}_{R_m}.$$

To prepare convergence factors R_is, we can see that

$$\begin{aligned} R_i &= 1 + \varepsilon^{2^i} = 1 + (1 - D_{i-1}) \\ &= 2 - D_{i-1}. \end{aligned}$$

It is actually the two's complement of $D_{i-1} = D \times R_0 \times R_1 \times \cdots R_{i-1}$. That is, R_i can be obtained by taking the 2's complement of D_{i-1} which is found in the previous iteration.

6.5.2 Divider Implementing Convergence Division Algorithm

The convergence division algorithm described previously has been implemented in several commercial computers including the IBM 360.

In the IBM 360, $D = 0.1 \times \times \cdots \times$ is a normalized floating-point fraction with 56 bits following the binary point. Here,

$$\frac{1}{2} \leq D < 1$$

$$\varepsilon = 1 - D \leq \frac{1}{2}$$

yield.

With a 56-bit precision the number 1.0 is represented as $0.11 \cdots 1$. According to the string property it is

$$1.00 \cdots 0\bar{1} = 1 - 2^{-56}.$$

That is, an error of 2^{-56} is incurred very possibly. Recall that
$$\varepsilon^{2^{(i+1)}} = 1 - D_i \leq \left(\frac{1}{2}\right)^{2^{(i+1)}}.$$
Suppose the algorithm stops when D_m is obtained, and we hope by then the difference between 1 and D_m to be
$$1 - D_m \leq 2^{-56}.$$
That is,
$$\left(\frac{1}{2}\right)^{2^{(m+1)}} \leq 2^{-56}.$$
To solve for m, we have
$$\begin{aligned}
-2^{(m+1)} &\leq -56 \\
2^{(m+1)} &\geq 56 \\
m &\geq \log_2 56 - 1 \\
m &= \lceil 5.8 - 1 \rceil = 5.
\end{aligned}$$

This implies that $m = 5$ is sufficient to generate Q:
$$Q = N \times R_0 \times R_1 \times R_2 \times R_3 \times R_4 \times R_5.$$
The division procedure in IBM 360 is as follows.

1. Get D and find the 2's complement of D (i.e., $R_0 = 2 - D$);
   $D_0 = D \times R_0 = D \times (2 - D)$;
   $N \leftarrow N \times R_0 = N \times (2 - D)$.

2. $R_1 = 2 - D_0$ (2's complement of D_0);
   $D_1 = D_0 \times R_1$; $N \leftarrow N \times R_1$.

3. $R_2 = 2 - D_1$;
   $D_2 = D_1 \times R_2$; $N \leftarrow N \times R_2$.

4. $R_3 = 2 - D_2$;
   $D_3 = D_2 \times R_3$; $N \leftarrow N \times R_3$.

5. $R_4 = 2 - D_3$;
   $D_4 = D_3 \times R_4$; $N \leftarrow N \times R_4$.

6. $R_5 = 2 - D_4$;
   $(D_5 = D_4 \times R_5;)$ $N \leftarrow N \times R_5$.

6.6 DIVISION BY DIVISOR RECIPROCATION

In this section another method for binary division using the multiplication approach is presented which can be implemented in an iterative way with simple logic circuits.

Let's rewrite the division expression as follows:

$$Q = \frac{N}{D} = N \times \left(\frac{1}{D}\right).$$

Here $\frac{1}{D}$ is referred to as the *Reciprocal of D*. Assume that D is a positive and normalized fraction, that is,

$$D > 0,$$

$$\frac{1}{2} \leq D < 1,$$

then

$$2 \geq \frac{1}{D} > 1. \qquad (6.6)$$

Also,

$$\frac{1}{r} \leq N$$

is assumed.

To find $\frac{1}{D}$, we cannot use "1 <u>divided</u> by D" since this goes back to the original problem − division. Alternatively a convergence approach is used to find $\frac{1}{D}$ iteratively and to complete the division by divisor reciprocation.

First, find an initial "approximation" to $\frac{1}{D}$ by table look up or combinational circuits. A ROM can be used for this purpose.

Let $p = \frac{1}{D}$, and the initial approximation of p be represented by p_0. We first make a selection for p_0, and then make it closer and closer to p iteratively.
Suppose that

$$D = 0.1 d_2 d_3 \cdots d_n$$

for

$$\frac{1}{2} \leq D < 1;$$

that is,

$$(0.10 \cdots 0)_2 \leq D < 1.$$

Investigate j bits following 0.1 in D,

$$D = 0.1 \underbrace{d_2 d_3 \cdots d_{j+1}}_{j\ bits} \cdots d_n.$$

Table 6.1: 2-Input 4-Output ROM to Store $p_0(s)$.

Input		Output			
d_2	d_3	x_1	x_2	x_3	x_4
0	0	1	1	0	0
0	1	0	1	1	1
1	0	0	0	1	1
1	1	0	0	0	1

The 2^j numbers, $0.1d_2d_3\cdots d_{j+1}$, will partition the range of D from $(0.10\cdots 0)_2$ to 1 into 2^j sub-ranges each having a length $\frac{1}{2}/2^j = 2^{-(j+1)}$. Let s be the indices of these sub-ranges. D must fall in one sub-range. For example if $j = 2$, [0.1<u>00</u>, 0.101), [0.1<u>01</u>, 0.110), [0.1<u>10</u>, 0.111) and [0.1<u>11</u>, 1.000) are the 4 sub-ranges. Given $D = 0.1\ \underbrace{10}_{j\ bits}\ 11111$, by examining the j (=2) bits following 0.1 we know that D falls in the 3rd sub-range where $s = 3$.

Given D in the sth sub-range ($s = 1, 2, \cdots, 2^j$), we find the midpoint in that subrange. The reciprocal of that midpoint value is used as the initial $\frac{1}{D}$ denoted as p_0.

The above procedure can be implemented by a ROM table. The j bits can be the input of the ROM. 2^j entries can be formed in the ROM table each corresponding to a sub-range mentioned before. As the reciprocal of the midpoint value of a sub-range p_0 is listed as an output in the ROM to provide the initial reciprocal approximation. A limited number of bits, say k bits are used to represent it, then k outputs are included in the ROM table. Since j and k are both comparatively smaller than n (the number of bits in D and $\frac{1}{D}$), the ROM implementation is cost effective. Note, since $1 < \frac{1}{D} \le 2$ by Equation (6.6), we always have

$$\frac{1}{D} = 1.x_1x_2\cdots x_n.$$

Integer 1 and the radix point are not necessary to be stored in ROM. Table 6.1 shows a ROM with input size $j = 2$ and output size $k = 4$. $x_1x_2x_3x_4$ represent only the fraction part of p_0. Given j and s, the formula shown below can find p_0, which needs to be calculated before stored in ROM.

$$p_0(s) = \frac{2^{j+1}}{2^j + s - \frac{1}{2}}.$$

This is because of the following reasons. Based on $\frac{1}{2}$, if D falls in sub-range s, then adding s multiples of the sub-range length and subtracting half of the sub-range length result in the value of the midpoint,

$$\frac{1}{2} + s \times 2^{-(j+1)} - \frac{1}{2} \times 2^{-(j+1)}.$$

The reciprocal of it is

$$\frac{1}{\frac{1}{2} + s \times 2^{-(j+1)} - \frac{1}{2} \times 2^{-(j+1)}}$$

$$= \frac{1}{\frac{1}{2} \times 2^{(j+1)} \times 2^{-(j+1)} + s \times 2^{-(j+1)} - \frac{1}{2} \times 2^{-(j+1)}}$$

$$= \frac{2^{(j+)}}{2^j + s - \frac{1}{2}}.$$

For example, given $s = 3$ the midpoint in the third sub-range has the value:

$$\frac{1}{2} + 3 \times 2^{-(j+1)} - \frac{1}{2} \times 2^{-(j+1)},$$

and the reciprocal of it is

$$p_0(s=3) = \frac{1}{\frac{1}{2} + 3 \times 2^{-(j+1)} - \frac{1}{2} \times 2^{-(j+1)}}$$

$$= \frac{2^{(j+1)}}{2^j + 3 - \frac{1}{2}}.$$

On the other hand, given $j = 2$,

$$p_0(s) = \frac{2^3}{2^2 + s - \frac{1}{2}}$$

$$= \frac{8}{4 + s - \frac{1}{2}}.$$

Hence we have

$$p_0(1) = \frac{8}{4 + 1 - \frac{1}{2}}$$

$$p_0(2) = \frac{8}{4 + 2 - \frac{1}{2}}$$

$$p_0(3) = \frac{8}{4 + 3 - \frac{1}{2}}$$

$$p_0(4) = \frac{8}{4 + 4 - \frac{1}{2}}.$$

See Figure 6.10 for illustration.

We can find that p_0 is a stepwise approximation to $\frac{1}{D}$, since p_0 is constant in each sub-range.

After p_0 is found set

$$a_0 = p_0 \times D.$$

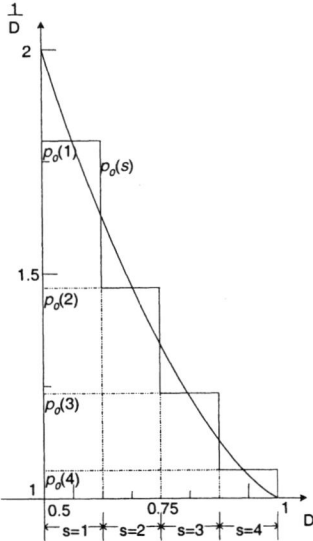

Fig. 6.10: Stepwise Approximation of the Reciprocal of Divisor

Then we recursively compute

$$p_i = p_{i-1} \times (2 - a_{i-1})$$
$$a_i = a_{i-1} \times (2 - a_{i-1})$$

with $i = 1, 2, \cdots$ being the recursion index.

Theoretically, as $i \to \infty$, we will have

$$\lim_{i \to \infty} p_i = \frac{1}{D}$$

$$\lim_{i \to \infty} a_i = 1.$$

To decide the number of iterations needed, say m, we recall that D is represented by 56 bits in IBM 360. An error of 2^{-56} will be incurred anyhow. Let ϵ denote such kind of error. After m iterations, there should be $p_m \approx \frac{1}{D}$, that is, $p_m \times D \approx 1$. Actually, $\mid 1 - (p_m \times D) \mid < \epsilon$ is sufficient since ϵ will be there anyway. So, when $i = m$ such that

$$\mid 1 - (p_m \times D) \mid < \epsilon,$$

the procedure should end where ϵ reflects the order of magnitude of the least significant bit and p_m is the final approximation of calculated reciprocal. Again, the actual number of iterations required is determined by the machine precision.

Let's look at an example of finding $\frac{1}{D}$. Suppose

$$D = (0.11011)_2,$$

that is,

$$D = (0.84375)_{10}.$$

$$\frac{1}{D} \approx 1.1851852$$

is the expected answer.

To find it, first find the initial approximation by

$$p_0(s) = \frac{2^{j+1}}{2^j + s - \frac{1}{2}}.$$

Let $j = 2$. We have $s = 3$ since $D = 0.1\underbrace{10}_{s=3}\cdots$. Recall that

$$d_2 d_3 = \begin{cases} 00 & \to & s = 1 \\ 01 & \to & s = 2 \\ 10 & \to & s = 3 \\ 11 & \to & s = 4 \end{cases}.$$

Hence

$$p_0(s) = \frac{2^3}{2^2 + 3 - \frac{1}{2}} = 1.2307692.$$

Second, the recursive process is done as follows:

$$\begin{aligned}
a_0 &= p_0 \times D = 1.2307692 \times 0.84375 \\
&= 1.0384615 \\
p_1 &= p_0 \times (2 - a_0) = 1.2307692 \times (2 - 1.0384615) \\
&= 1.183432
\end{aligned}$$

$$\begin{aligned}
a_1 &= a_0 \times (2 - a_0) \\
&= 1.0384615 \times (2 - 1.0384615) \\
&= 0.9985207 \\
p_2 &= p_1 \times (2 - a_1) = 1.183432 \times (2 - 0.9985207) \\
&= 1.1851826
\end{aligned}$$

$$a_2 = a_1 \times (2 - a_1)$$

162 SEQUENTIAL DIVISION

$$\begin{aligned} &= 0.9985207 \times (2 - 0.9985207) \\ &= 0.9999978 \\ p_3 &= p_2 \times (2 - a_2) = 1.1851826 \times (2 - 0.9999978) \\ &= 1.1851852. \end{aligned}$$

$\vdots$

Depending on ϵ, the process stops when

$$\mid 1 - (p_m \times D) \mid < \epsilon.$$

In the above example

$$\mid 1 - (p_3 \times D) \mid = \mid 1 - (1.1851852 \times 1.84375) \mid \equiv 0.$$

REFERENCES

1. F. S. Anderson, et al., "The IBM System/360 Model 91: Floating-Point Execution Unit," *IBM Journal Res. and Dev.*, 11 (Jan. 1967), pp. 34–53.

2. D. E. Atkins, "Higher-Radix Division Using Estimates of the Divisor and Partial Remainders," *IEEE Trans. Comp.*, Vol. C-17, No. 10, Oct. 1968, pp. 925–934.

3. D. E. Atkins, "The Theory and Implementation of SRT Division," Tech. Report, No. 230, Dept. of Computer Science, University of Illinois, Urbana, 1967.

4. D. E. Atkins, "Design of Arithmetic Units of ILLIACIII: Use of Redundancy and Higher Radix Methods," *IEEE Trans. Comp.*, Aug. 1970, pp. 720–733.

5. A. Avizienis, "The Recursive Division Algorithm," Class Notes, Engr. 225A, UCLA, Los Angeles, 1971.

6. W. Buchholz, ed., *Planning a Computer System*, McGraw-Hill, New York, 1962, pp. 214–216.

7. A. W. Burks, H. H. Goldstine and J. von Neumann, "Preliminary Discussion of the Logical Design of an Electronic Computing Instrument," Institute for Advanced Study, Princeton, N. J., 1946 (reprinted in C. G. Bell and A. Newell, *Computer Structures Readings and Examples*, McGraw-Hill, New York, 1971).

8. D. Ferrari, "A Division Method Using a Parallel Multiplier," *IEEE Trans. Comp.*, Ec-16, (Apr. 1967), pp. 224–226.

9. M. J. Flynn, "On Division by Functional Iteration," *IEEE Trans. Comp.*, C-19 (Aug. 1970), 702–706.

10. D. L. Fowler and J. E. Smith, "An Accurate High Speed Implementation of Division by Reciprocal Approximation," *Proc. 9th Symp. on Comp. Arith.* (Sept. 1989), pp. 60–67.

11. C. W. Freiman, "Statistical Analysis of Certain Binary Division Algorithms," *Proc. IRE*, Vol. 49, No. 1, Jan. 1961, pp. 91–103.

12. H. L. Garner, "Number Systems and Arithmetic," in *Advances in Computers*, Vol. 6, F. L. Alt and M. Rubinoff, eds. Academic Press, New York, 1965. pp. 168–177.

13. R. E. Gilman, "A Mathematical Procedure for Machine Division," *Communication of Assoc. for Comp. Mach.*, Vol. C, No. 4, Apr. 1959, pp. 10–12.

14. D. Goldberg, D. A. Patterson and J. L. Hennessy, Computer Arithmetic, in *Computer Architecture: A Quantiative Approach*, Morgan Kaufmann, San Mateo, CA, 1996.

15. R. Z. Goldschmidt, "Applications of Division by Convergence," M.S. Thesis, M.I.T. Cambridge, MA, June 1964.

16. D. R. Hartree, *Calculating Instruments and Machines*, University of Illinois Press, Urbana, 1949, p. 57.

17. V. U. Kalaycioglu, "Analysis and Synthesis of Generalized-Radix Additive Normalization Division Techniques," Tech. Report, Dept. of Elec. & Comp. Engr., University of Michigan, Ann Arbor, May 1975.

18. E. V. Krishnamurthy, "On Range-Transformation Techniques for Division," *IEEE Trans. Comp.*, Vol. C-19, No. 3, Mar. 1970, pp. 227–231.

19. O. L. MacSorley, "High-Speed Arithmetic in Binary Computers," *Proc. IRE*, Vol. 49, Jan. 1961, pp. 67–91.

20. P. W. Markstein, "Computation of Elementary Functions on the IBM RISC System/6000 Processor," *IBM Journal Res. and Dev.*, 34 (Jan. 1990), pp. 111–119.

21. G. Metze, "A Class of Binary Divisions Yielding Minimally Represented Quotients," *IRE Trans.* Vol. EC-11, No. 6, Dec. 1962, pp. 761–764.

22. J. E. Robertson, "A New Class of Digital Division Methods," *IEEE Trans. Comp.*, Vol. C-7, Sept. 1958, pp. 218–222.

23. J. E. Robertson, "The Correspondence Between Methods of Digital Division and Multiplier Recoding Procedures," *IEEE Trans. Comp.*, Vol. C-19, No. 8, Aug. 1970, pp. 692–701.

24. R. R. Shively, "Stationary Distribution of Partial Remainders in SRT Digital Division," Ph.D. Thesis, University of Illinois, Urbana, 1963.

25. G. S Taylor, "Radix 16 SRT Dividers with Overlapped Quotient Selection Stages," *Proc. 7th Symp. on Computer Arithmetic* (June 1985), pp. 64–71.

26. K. D. Tocher, "Techniques of Multiplication and Division for Automatic Binary Computers," *Quart. J. Mech. Appl. Math.*, Vol. XI, Pt. 3, 1958, pp. 364–384.

27. C. Tung, "A Division Algorithm for Signed-Digit Arithmetic," *IEEE Trans. Comp.*, Vol. 17, 1968, pp. 887–889.

28. C. Tung, "Arithmetic," Chap. 3 in *Computer Science*, A. F. Cardenas, et al. eds., Wiley Interscience, New York, 1972.

29. J. B Wilson, et al, "An Algorithm for Rapid Binary Division," *IRE Tran.* Vol. EC-10, No. 4, Dec. 1961, pp. 662–670.

PROBLEMS

6.1 Perform the following binary Nonrestoring Division.

$$+0.011 \overline{\smash{\big)}\ +0.001011}$$

6.2 Find $A = (10110101)_2$ divided by $B = (1101)_2$ applying
(a) Binary Restoring algorithm;
(b) Binary Non-restoring algorithm.

6.3 Write a program to convert $Q = (\bar{7}8\bar{4}0\bar{1}2\bar{3})_{10}$ into a 10's complement number. Print out the results of q'_j and q'_{j+1} in each iteration.

6.4 Show the numerical procedure of $A = (0.00100011)_2$ divided by $B = (0.0101)_2$ applying Wilson-Ledley's division algorithm. Indicate in the flow chart of Figure 6.7 the actual steps taken for each iteration in this division.

6.5 Apply Wilson-Ledley's algorithm to find an 8-bit quotient $Q = \frac{N}{D}$ for $N = (0.01101011)_2$ and $D = (0.1001)_2$.

6.6 Use Convergence Division methodologies to perform the same division as in Problem 6.4. Show the procedure of this division referencing the IBM 360 procedure.

6.7 In the convergence procedure $D_i = D_{i-1} \cdot R_i$ where $D_0 = D \cdot R_0$ and $D = 1 - \delta$, R_i can choose such that

$$D_0 = D \cdot R_0 = (1-\delta)(1+\delta+\delta^2) = 1-\delta^3$$
$$D_0 = D \cdot R_0 = (1-\delta)(1+\delta+\delta^2) = 1-\delta^3,$$
$$\ldots$$

where $R_i = 1 + \delta^{3^i} + \delta^{2 \cdot 3^i}$ and $D_{i-1} = 1 - \delta^{3^i}$. Compare such cubic convergence division with the quadratic convergence division presented in the text in terms of the convergence rate and cycle time.

6.8 Find the reciprocal of $D = (0.11101)_2$ by the way of stepwise approximation, and let $j = 3$ and $\epsilon = 0.0001$. How close is the initial reciprocal of D to the real reciprocal?

6.9 Consider the division $Q = N/D$ through the Divisor Reciprocation.
a) If in each of the 2^j equal intervals between $\frac{1}{2}$ and 1, a leftmost point (instead of a midpoint) is chosen for the initial approximation. Find a formula to express the initial reciprocal of divisor in the hth interval, $p_0(h)$, as a function of j and h.
b) According to your formula, build a ROM table with 2 inputs and 4 outputs.
c) Given $D = 0.1000$, find the initial reciprocal of it from your ROM table.
d) Find the final reciprocal of D by the recursive process. (Show your operation in binary and allow error $< 2^{-10}$.)

6.10 Given $D = (0.11011)_2$, use the ROM table in the text to obtain the reciprocal of D represented by a 10-bit binary fraction. Please show your operation in binary and pay attention to when to stop.

7
Fast Array Dividers

In this chapter, the design and construction of various high-speed iterative cellular arrays for parallel divisions are discussed. In the proposed array dividers, a large amount of replicated units are used for the comparison of the partial remainder and divisor, and the shift is realized by physical wiring. Three types of array dividers are to be introduced: (1) the Restoring Array Dividers, (2) the Non-restoring Array Dividers, and (3) the Carry-Lookahead Array Dividers. In addition, their performance and cost-effectiveness are to be analyzed.

7.1 RESTORING CELLULAR ARRAY DIVIDER

The restoring cellular array divider is based on the "restoring" division algorithm. Recall the circuit schematic of the divider based on the restoring division method shown in Figure 6.5. Partial remainders are stored in a register and subtractions then take place in a two-operand adder. Restorations are realized by the MUX. Hardware for this approach is simple but slow. With the restoring cellular array divider presented below, the execution can be made much faster.

Let dividend $A = .a_1 a_2 \cdots a_{2n}$, divisor $D = .d_1 d_2 \cdots d_n$ and quotient $Q = .q_1 q_2 \cdots q_n$. Figure 7.1 shows a schematic logic diagram of an n-by-n restoring array divider with $n = 4$.

The basic element shown in the figure is a controlled subtracter (CS) cell, in which $a - d$ is performed if mode $P = 0$, to find the difference between the previous partial remainder and the divisor. The borrow signal

$$c_{out} = \bar{a}d + \bar{a}c_{in} + dc_{in},$$

168 FAST ARRAY DIVIDERS

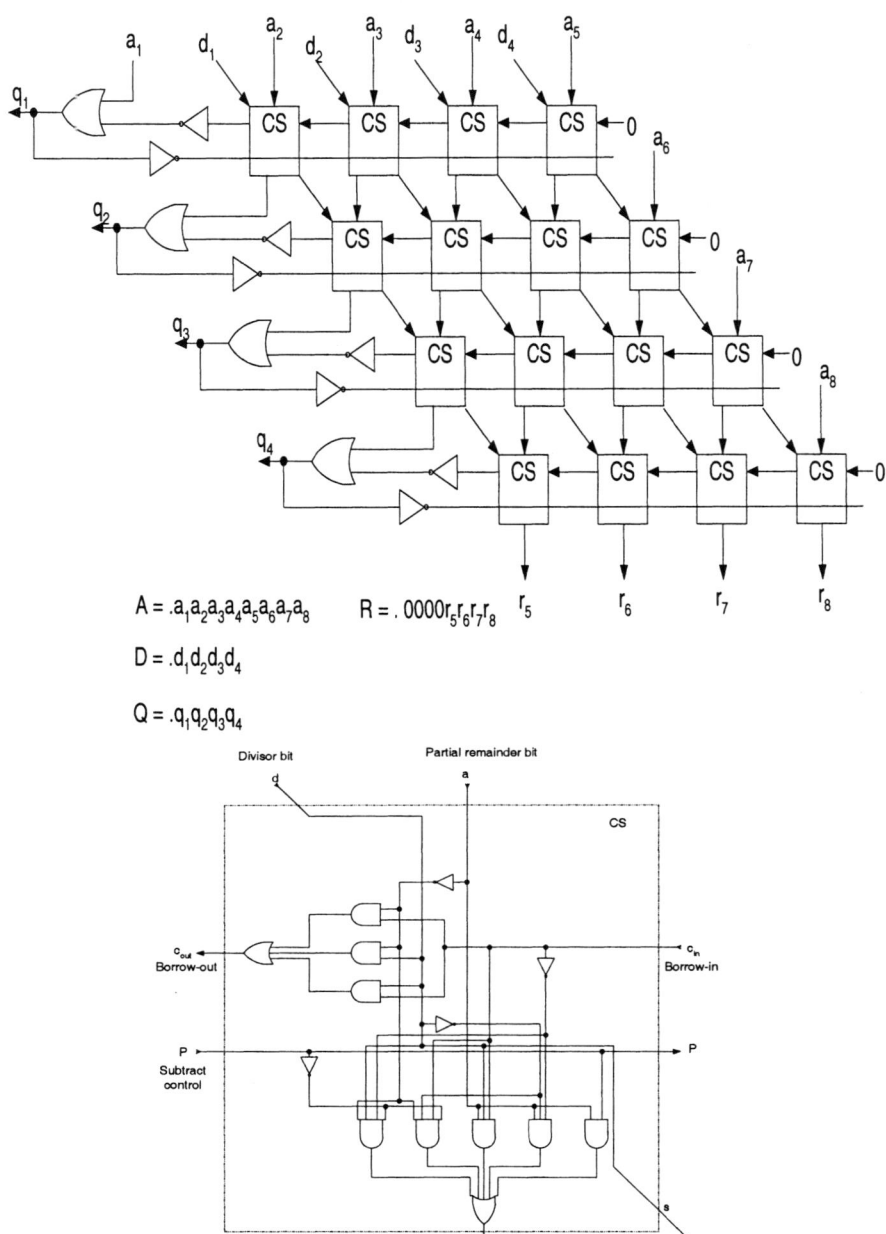

$A = .a_1a_2a_3a_4a_5a_6a_7a_8$ $R = .0000r_5r_6r_7r_8$

$D = .d_1d_2d_3d_4$

$Q = .q_1q_2q_3q_4$

Fig. 7.1: 4-by-4 Restoring Array Divider

and the difference

$$s = aP + a\bar{d}\bar{c}_{in} + adc_{in} + \bar{a}\bar{d}c_{in}\bar{P} + \bar{a}dc_{in}\bar{P},$$

that is,

$$s = \begin{cases} a \oplus d \oplus c_{in} & \text{if } P = 0 \text{ (subtract)}, \\ a & \text{if } P = 1 \text{ (no operation)}. \end{cases}$$

Instead of shifting the partial remainder left to form $rR^{(j)}$, equivalently we have the remainder fixed and shift the divisor right along the diagonal lines. q_is are obtained from the left of each row.

$$q_i = a_i + \bar{c}_{out},$$

where a_i is the bit shifted out in the $(i-1)$th iteration. When $a_i = 1$, which means $rR^{(j)} > D$, the subtraction is successful. Or, if c' is the borrow-out signal in the leftmost position of each row, when $\bar{c}' = 1$, that is, $c' = 0$ meaning no borrow, the subtraction is successful.

P is the control signal.

$$P = \begin{cases} 0 & \text{if performing subtraction}, \\ 1 & \text{no operation}. \end{cases}$$

Here,

$$P = \bar{q}_i.$$

$P = 0$, $q_i = 1$ are for the subtraction successful case, while $P = 1$, $q_i = 0$ are for the no operation case. Note that in the restoring division, we need to add the divisor back after an unsuccessful subtraction. These two operations canceled each other, that is, no operation is to be performed.

The final remainder is represented by $0.0000r_5r_6r_7r_8$ in the above example. In general, an $n \times n$ restoring binary cellular array divider receives a $2n$-bit dividend and n-bit divisor, and produces an n-bit quotient and $2n$-bit remainder including n leading 0s.

Note that the AND-OR logic relationship can be realized by NAND-NAND gates, and the following logic function

$$c_{out} = \bar{a}d + \bar{a}c_{in} + bc_{in},$$

$$s = aP + a\bar{d}\bar{c}_{in} + adc_{in} + \bar{a}\bar{d}c_{in}\bar{P} + \bar{a}dc_{in}\bar{P},$$

can be completed within $3\triangle_g$ time. For n-bit division, an array of size $n \times n$ is required.

In a CS cell, the signal propagation from input a to output c_{out} needs $3\triangle_g$ time if c_{in} is available. For all the CS cells in a row, such propagation can be completed in parallel. The signal propagation from input c_{in} to output c_{out} takes $2\triangle_g$. For all the cells in a row, except that in the rightmost cell such delay can be overlapped with the delay from a to c_{out}, the delay times in the other $(n-1)$ cells accumulate. That makes a total delay of $2(n-1)\triangle_g$. It takes $4\triangle_g$ for c' to be fed back into P on the left of each row. After P is ready, $3\triangle_g$ are required to obtain output s, which is a parallel delay time for all the cells in a row. Hence, for the first row, the total delay time is

$$[3 + 2(n-1) + 4 + 3]\triangle_g = (2n + 8)\triangle_g.$$

For the second row and every row below, the rightmost cell is not dependent on the s from the above row. The c_{out} signal in that cell can be ready early, overlapping the delay time with that required by the upper row operations. For the 2nd cell from right and $(n-2)$ cells on its left, $3\triangle_g$ parallel delay time is required for the propagation from s (or a) to c_{out}. After that, $2(n-2)\triangle_g$ accumulated delay time is required for c_{in} to c_{out} propagation toward the left of the row. Including the $4\triangle_g$ for feeding back D, and $3\triangle_g$ for P to s propagation,

$$[3 + 2(n-2) + 4 + 3]\triangle_g = 2n + 6\triangle_g$$

time is required for each row other than the first. There are $(n-1)$ such rows, hence the total delay for which is

$$[(2n+6) \times (n-1)]\triangle_g = 2n^2 + 4n - 6\triangle_g.$$

Adding on the delay for the first row, we have the total delay time for the restoring cellular array divider as

$$[(2n+8) + (2n^2 + 4n - 6)]\triangle_g = 2n^2 + 6n + 2\triangle_g.$$

The total number of controlled subtracters (CS) required is n^2. In each of the CS cells 14 gates are counted. Taking into consideration the one OR gate and two NOT gates on the very left of each row, there is an extra gate count of $3n$ yields. All together the gate count of the restoring array divider is

$$14n^2 + 3n.$$

Realizing the AND-OR function by NAND-NAND gates each requiring area $1A_g$, we have the area complexity of the restoring array divider calculated as follows:

$$\begin{aligned} & 14(n \times n)A_g + (A_{OR} + 2A_{NOT}) \times n \\ = & 14n \times nA_G + (2+2)nA_g \\ = & (14n^2 + 4n)A_g. \end{aligned}$$

7.2 NON-RESTORING CELLULAR ARRAY DIVIDER

Recall that in the non-restoring division, restoration is not required. Rather, the only operation is either addition or subtraction. Successively right-shifted versions of the divisor are subtracted from or added to the dividend, resulting in partial remainders. The sign of the partial remainder determines the quotient bit and, further, determines whether to add or subtract the shifted divisor in the next cycle. Note that if 2's complement arithmetic is utilized, the addition/subtraction can be easily handled and implemented by hardware.

Let dividend $A = a_0.a_1a_2 \cdots a_{2n}$, divisor $D = d_0.d_1d_2 \cdots d_n$ and quotient $Q = q_0.q_1q_2 \cdots q_n$. The operands are assumed to be positive, normalized fractions, so that $a_0 = d_0 = 0$ and $a_1 = d_1 = 1$. Since $\frac{1}{2} \leq (A, D) < 1$, $A_{min}/D_{max} < Q < A_{max}/D_{min}$, hence $\frac{1}{2} < Q < 2$. To perform the n-bit parallel non-restoring division, an $(n+1)$-by-$(n+1)$ non-restoring array divider is required. Figure 7.2 shows the schematic logic diagram of a 5-by-5 non-restoring array divider, where $n + 1 = 5$ and 4-bit (excluding bit 0) division is performed. Note that q_0, a_0 and d_0 are involved in the operation here. In contrast, in the previously described restoring array divider, they are not used for calculation. Also, in the array divider described in this section, Q is not a signed-digit number anymore. It is a 2's complement number.

The non-restoring array divider consists of rows of carry-propagate adders with each logic cell containing a full adder and an exclusive-OR gate. Again the partial remainder is fixed and the divisor shifts right bit by bit. The exclusive-OR gate controls the divisor input to the full adder. The control signal P determines whether an addition or subtraction is to be performed. Subtraction is performed in 2's complement form by forming the 1's complement of the divisor and forcing a carry into the rightmost cell.

The carry-out signal of the leftmost cell is actually the sign of the partial remainder. Quotient digit q_i is dependant on it. If the partial remainder < 0, $q_i = 0$. If the partial remainder > 0, $q_i = 1$. Also, the quotient bit obtained in the previous iteration (upper row) is used as the control signal P for addition/subtraction selection in the next iteration (lower row). If $P = 0$, an addition is to be performed. If $P = 1$, a subtraction is to be performed by adding a 2's complement number. Note that since the partial remainder and the multiple of divisor always have opposite signs, in the leftmost cell two operands of the FA add up to 1. $c_{out} = 1$ only if the third operand is 1, in which case the sum is 0 indicating that the new partial remainder has a positive sign.

The sum (partial remainder) can be obtained as follows:

$$\begin{aligned} s_i &= a_i \oplus (d_i \oplus P) \oplus c_i \\ &= \bar{a}_id_ic_iP + a_i\bar{d}_ic_iP + a_id_i\bar{c}_iP + a_id_ic_i\bar{P} + a_i\bar{d}_i\bar{c}_i\bar{P} + \bar{a}_id_i\bar{c}_i\bar{P} \\ &\quad + \bar{a}_i\bar{d}_ic_i\bar{P} + \bar{a}_i\bar{d}_i\bar{c}_iP. \end{aligned}$$

For 3-level gates, $3\triangle_g$ delay time is required,

$$c_{i+1} = (a_i + c_i)(d_i \oplus P) + a_ic_i$$

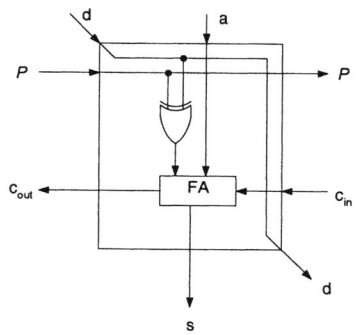

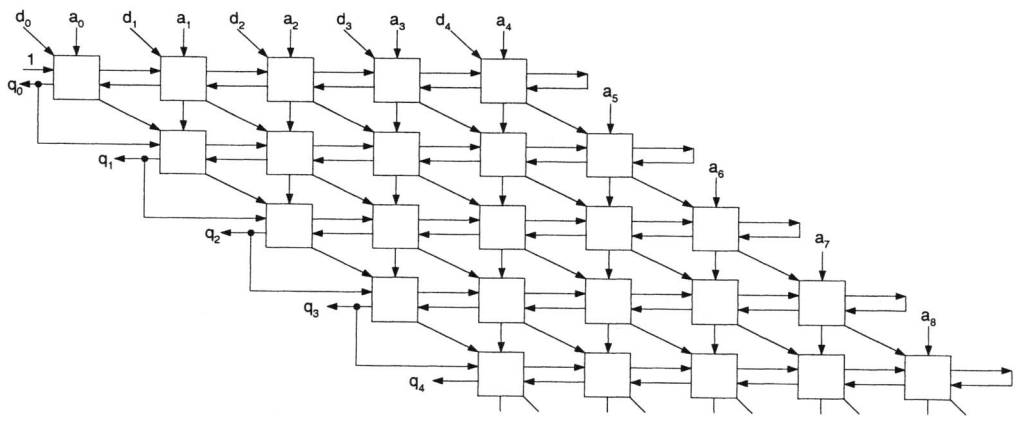

$A = a_0.a_1a_2a_3a_4a_5a_6a_7a_8$
$D = d_0.d_1d_2d_3d_4$
$Q = q_0.q_1q_2q_3q_4$
$R = 0.0000r_5r_6r_7r_8$

Fig. 7.2: 5-by-5 Non-Restoring Array Divider

$$= a_i d_i \bar{P} + a_i \bar{d}_i P + d_i c_i \bar{P} + \bar{d}_i c_i P + a_i c_i.$$

An $(n+1) \times (n+1)$ non-restoring array divider requires $(n+1)^2$ Controlled Add/Subtract Cells (CASs).

For n-bit division applying the non-restoring array divider, the required array size is $(n+1) \times (n+1)$. The delay time for one row is

$$\triangle_{XOR} + \underbrace{\triangle_{FA} + n \cdot \triangle_{carry-propag}}_{\text{or } (n+1)\triangle_{carry-propag}}$$
$$= (2 + 2 + n \cdot 2)\triangle_g$$
$$= (2n + 4)\triangle_g.$$

For the FA in the hard-wired version, C and S are assumed to be generated at the same time.

There are $(n+1)$ rows in total. Only if the q_i from the above row is obtained, the next row can start operating. Hence the total delay time is

$$(2n + 4) \times (n + 1) = (2n^2 + 6n + 4)\triangle_g.$$

The gate count for the non-restoring array divider can be decided as follows. Assume that both the original variables and the complement of them are available. The sum of 8 products in the logic expression for s_i,

$$s_i = \bar{a}_i d_i c_i P + a_i \bar{d}_i c_i P + a_i d_i \bar{c}_i P + a_i d_i c_i \bar{P} + a_i \bar{d}_i \bar{c}_i \bar{P} + \bar{a}_i d_i \bar{c}_i \bar{P}$$
$$+ \bar{a}_i \bar{d}_i c_i \bar{P} + \bar{a}_i \bar{d}_i \bar{c}_i P$$

can be realized by 8 NAND gates in the first level and 1 NAND gate in the second level. Similarly, the sum of 5 products in the logic expression for c_{i+1},

$$c_{i+1} = a_i d_i \bar{P} + a_i \bar{d}_i P + d_i c_i \bar{P} + \bar{d}_i c_i P + a_i c_i$$

can be realized by 5 NAND gates in the first level and 1 NAND gate in the second level. Hence, the total gate count of a non-restoring array divider is

$$8 + 1 + 5 + 1 = 15.$$

7.3 CARRY-LOOKAHEAD CELLULAR ARRAY DIVIDER

The previous dividers we discussed always involve a carry (borrow) propagation along the row. The length of the row is proportional to the number of bits, and the delay time becomes very considerable for a long row. The carry-out in each row determines the sign bit of the partial remainder which selects the quotient bit and the add/subtract operation in the next row. Carry-lookahead circuitry can be included for the best possible solution from the speed point of view. The array is presented in Figure 7.3 with the inputs

$$A = A_0.A_1 A_2 \cdots A_n,$$

complemented

$$D = D_0.D_1 D_2 \cdots D_n,$$

and output

$$Q = Q_0.Q_1 Q_2 \cdots Q_n.$$

There are A cells, S cells and CLA cells in the array. The A_js (dividend) are input from the top, and the complemented D_js (divisor) are input through the diagonal lines. Addition or subtraction is to be performed in the A cells according to the same policy introduced in the non-restoring division approach (with an S cell at the very left of each row). Each cell is actually a controlled carry-save adder/subtractor, and two outputs are to be generated, a sum and a carry. S_j^i is the sum bit generated in the jth bit position of ith row, and C_j^i and C_{j-1}^i are the carry-in and carry-out of that position, respectively. The two outputs resulting from the upper row are input to the row below. Representing them in vectors we have

$$
\begin{array}{rcccccccl}
S^{i-1} & = & S_0^{i-1}. & S_1^{i-1} & S_2^{i-1} & \cdots & S_{n-1}^{i-1} & A_{n+i} & \text{sum vector} \\
C^{i-1} & = & C_0^{i-1}. & C_1^{i-1} & C_2^{i-1} & \cdots & 0 & Q_{i-1} & \text{carry vector} \\
\pm D & = & D_0. & D_1 & D_2 & \cdots & D_{n-1} & D_n & \\
\hline
S^i & = & S_0^i. & S_1^i & S_2^i & \cdots & S_{n-1}^i & S_n^i & \\
C^i & = & C_0^i. & C_1^i & C_2^i & \cdots & C_{n-1}^i & 0. & \\
\end{array}
$$

Each A cell implements

$$S_j^i = S_j^{i-1} \oplus C_j^{i-1} \oplus (D_j \oplus K^i)$$
$$C_{j-1}^i = (D_j \oplus K^i)(S_j^{i-1} + C_j^{i-1}) + S_j^{i-1} C_j^{i-1}.$$

The S cell combines the very left bits S_0^i and C_0^i in the two output vectors by implementing

$$s_0^i = S_0^{i-1} \oplus C_0^{i-1} \oplus \bar{K}^i) \oplus C_0^i, \qquad (7.1)$$

where $\bar{K}^i$ is Q_{j-1}. When $Q_{j-1} = 0$, addition is to be perfromed and D_0 is to be added. $D_0 = 0 = \bar{K}^i$. When $Q_{j-1} = 1$, subtraction is to be performed and $\bar{D}_0$ is to be added. $\bar{D}_0 = 1 = \bar{K}^i$. So, $\bar{K}^i$ is XORed in Equation (7.1). The last term C_0^i is only the carry generated by bit position 1, an isolated carry information, as well as $C_1^i, C_2^i, \cdots, C_{n-1}^i$. Whether they will propagate to bit position 0 and affect the sign can be told by carry-lookahead.

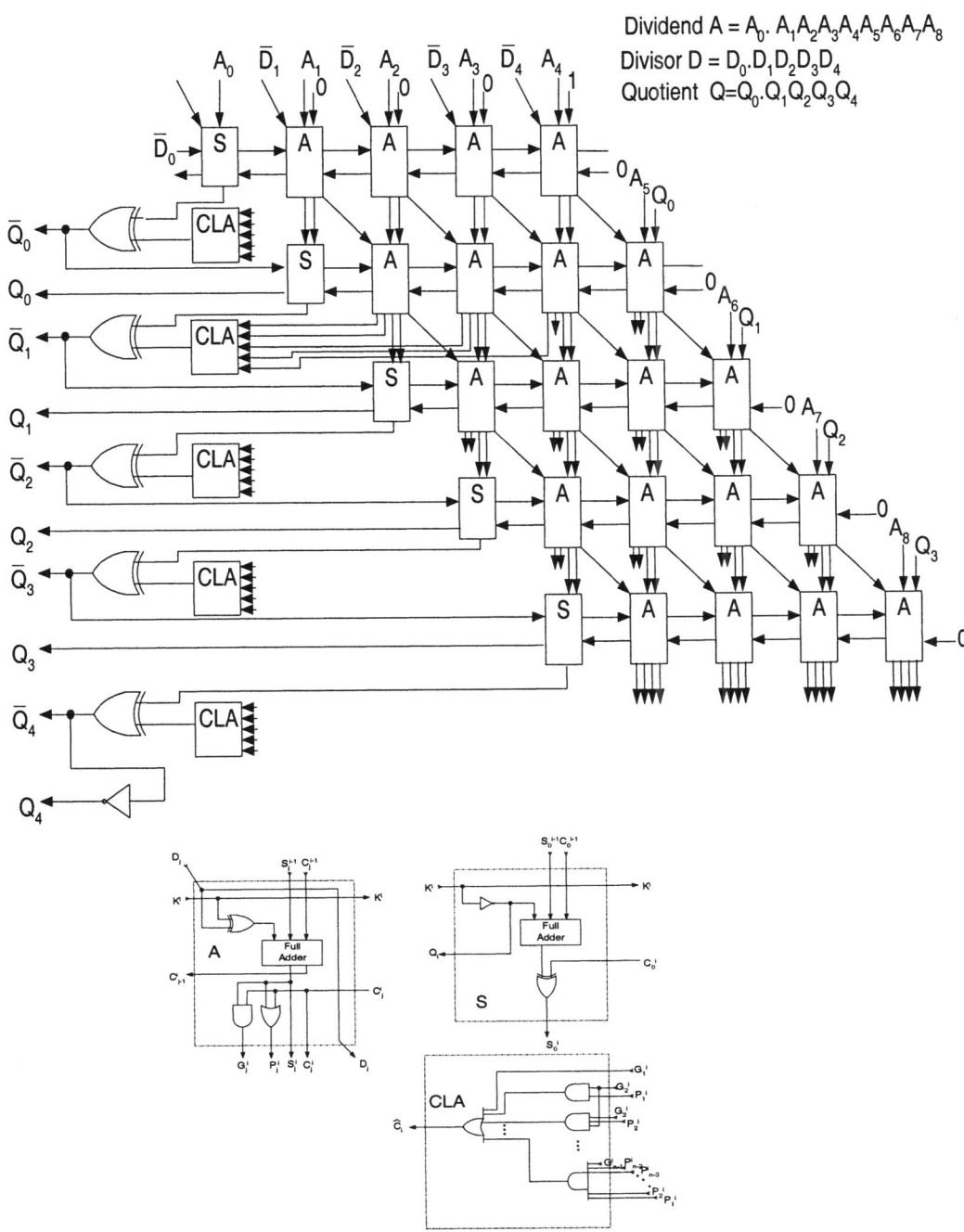

Fig. 7.3: Carry-Lookahead Array Divider for 4-bit Division
(Carry-Lookahead Mechanism is Shown in the Second Row Only)

176 FAST ARRAY DIVIDERS

The CLA cells are for the carry-lookahead. Let

$$G_j^i = C_j^i S_j^i$$
$$P_j^i = C_j^i \oplus S_j^i.$$

The output of carry-lookahead is

$$\hat{C}^i = G_1^i + P_1^i G_2^i + P_1^i P_2^i G_3^i + \cdots \\ + P_1^i P_2^i \cdots P_{n-2}^i G_{n-1}^i.$$

K^i is a control signal propagating from left to right in the ith row without any delay. Since the carry is saved, we do not wait on it to propagate from the rightmost cell to the leftmost cell. So the operation carried here is independent of the word length n. Hence the delay time of the carry-lookahead array divider is

$$\triangle_{CLA} = (n+1) \times \triangle_{row} + \triangle_{not}.$$

We can see that the total delay time is $O(n)$. In other words, it is linear to n. Recall that for the restoring and non-restoring divisions the time needed is $O(n^2)$.

Particularly, all the operations in A cells are done in parallel with the delay time starting from whenever K is ready, that is,

$$\triangle_{XOR} + \triangle_{FA} = (2+2)\triangle_g = 4\triangle_g.$$

Then, two more $\triangle_g$ are required to obtain G_j^i and P_j^i, which can be overlapped with the delay of the XOR gate in S cell. After that, $2\triangle_g$ are required by the NAND-NAND function to obtain $\hat{C}^i$ in CLA. Within two more $\triangle_g$, K^i can be fed to the next row. So,

$$\triangle_{row} = 4 + 2 + 2 + 2 = 10\triangle_g.$$

Since there are $(n+1)$ rows in total and one extra inverter in the bottom row, the total delay time for the array is

$$\triangle_{array} = (n+1) \times \triangle_{row} + \triangle_{not} \\ = ((n+1) \times 10 + 1)\triangle_g \\ = (10n + 11)\triangle_g.$$

The area complexity for the carry-lookahead array divider can be analyzed as follows. To perform an n-bit division we need $n(n+1)$ A cells, $(n+1)$ S cells, $(n+1)$ CLA circuits, $(n+1)$ XOR gates and one inverter.

In A cell,

$$S_j^i = S_j^{i-1} \oplus C_j^{i-1} \oplus (D_j \oplus K^i).$$

That is,

$$S_j^i = \bar{S}_j^{i-1} \bar{C}_j^{i-1} \bar{D}_j K^i + \bar{S}_j^{i-1} \bar{C}_j^{i-1} D_j \bar{K}^i + \bar{S}_j^{i-1} C^{i-1}{}_j \bar{D}_j \bar{K}^i + \bar{S}_j^{i-1} C^{i-1}{}_j D_j K^i$$

$$+ S_j^{i-1}\bar{C}^{i-1}{}_j\bar{D}_j\bar{K}^i + S_j^{i-1}\bar{C}^{i-1}{}_j D_j K^i + S_j^{i-1}C_j^{i-1}D_j\bar{K}^i + S_j^{i-1}C_j^{i-1}\bar{D}_j K^i.$$

Here, 9 gates are required assuming that each variable and its complement are both available.

$$C_{j-1}^i = (D_j \oplus K^i)(S_j^{i-1} + C_j^{i-1}) + S_j^{i-1}C_j^{i-1}.$$

That is,

$$C_{j-1}^i = \bar{D}_j K^i S_j^{i-1} + \bar{D}_j K^i C_j^{i-1} + D_j \bar{K}^i S_j^{i-1} + D_j \bar{K}^i C_j^{i-1} + S_j^{i-1}C_j^{i-1}.$$

Here 6 gates are required. Taking into consideration the 2 gates realizing the following functions,

$$G_j^i = C_j^i S_j^i$$
$$P_j^i = C_j^i \oplus S_j^i,$$

17 gates are required for each A cell.

In S cell,

$$S_0^i = (S_0^{i-1} \oplus C_0^{i-1} \oplus \bar{K}^i) \oplus C_0^i.$$

That is,

$$\begin{aligned}S_0^i &= \bar{S}_0^{i-1}C_0^{i-1}K^i\bar{C}_0^i + \bar{S}_0^{i-1}C_0^{i-1}\bar{K}^iC_0^i + S_0^{i-1}\bar{C}_0^{i-1}K^i\bar{C}_0^i + S_0^{i-1}\bar{C}_0^{i-1}\bar{K}^iC_0^i \\ &+ \bar{S}_0^{i-1}\bar{C}_0^{i-1}K^iC_0^i + \bar{S}_0^{i-1}\bar{C}_0^{i-1}\bar{K}^i\bar{C}_0^i + S_0^{i-1}C_0^{i-1}K^iC_0^i + S_0^{i-1}C_0^{i-1}\bar{K}^i\bar{C}_0^i,\end{aligned}$$

and 9 gates are required to implement an S cell.

For a CLA array divider performing n-bit division, the number of gates required per row is

$$\underbrace{17n}_{A\ cells} + \underbrace{9}_{S\ cell} + \underbrace{(n-1)}_{CLA\ cell} + \underbrace{1}_{left\ most\ XOR}$$
$$= 18n + 9.$$

There are in total $(n+1)$ rows,

$$(18n + 9) \times (n+1) + 1 = 18n^2 + 27n + 10$$

gates are required including the NOT gate in the bottom row.

178   FAST ARRAY DIVIDERS

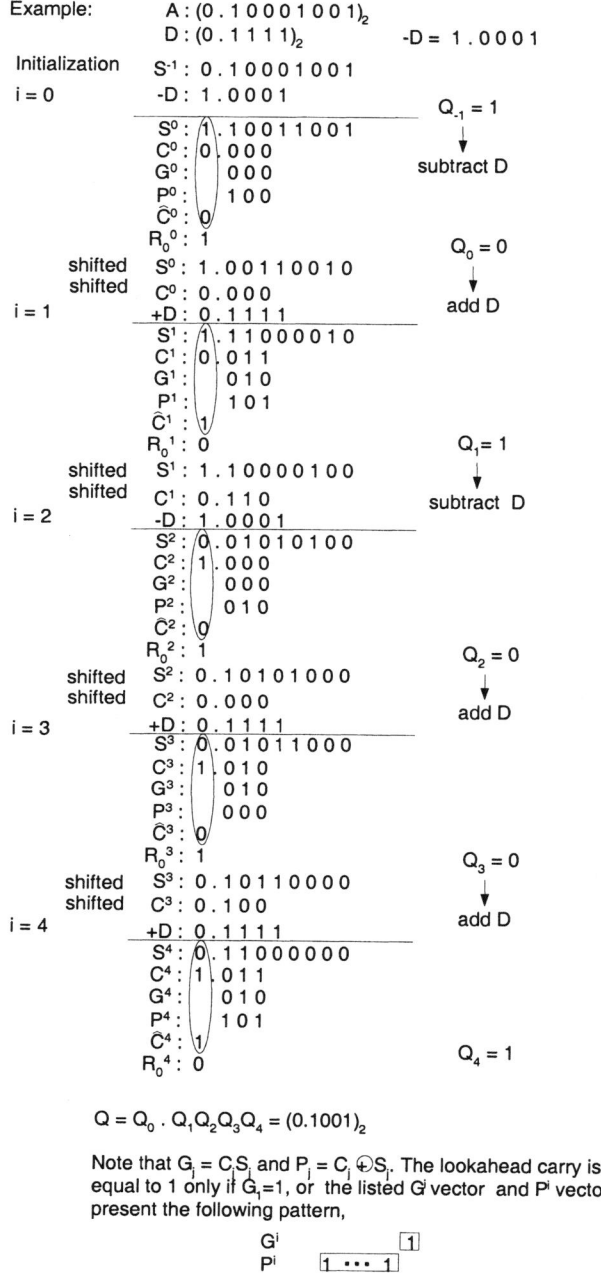

Fig. 7.4: Example of Carry-Lookahead Array Division

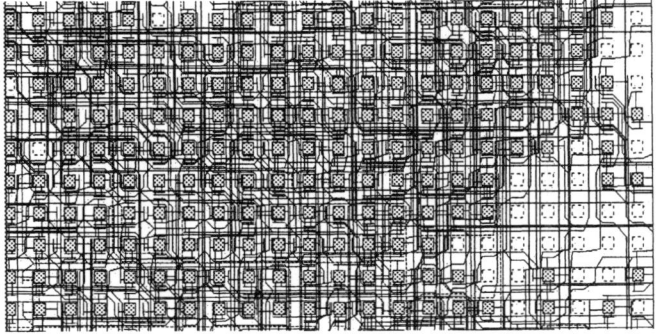

Fig. 7.5: Wires Can Take Up Significant Space

The algorithm performed by the carry-lookahead array divider is as follows.

Initialization:
$S^{-1} = A(dividend)$; $C^{-1} = 00\cdots 1$, and $Q_{-1} = 1$
Do for $i = 0$ to n.
Step 1. Generate the two-vectors, carry and partial remainder:
  if $Q_{i-1} = 1$, $C^i \circ S^i \leftarrow S^{i-1} + C^{i-1} - D$;
  if $Q_{i-1} = 0$, $C^i \circ S^i \leftarrow S^{i-1} + C^{i-1} + D$.
Step 2. Lookahead: $\hat{C}^i = G_1^i + P_1^i G_2^i + P_1^i P_2^i G_3^i + \cdots + P_1^i \cdots P_{n-2}^i G_{n-1}^i$
  where $G_j^i = S_j^i C_j^i$ and $P_j^i = S_j^i \oplus C_j^i$.
Step 3. Decide the sign for partial remainder:
  $R_0 = S_0^i \oplus C_0^i \oplus \hat{C}^i$, and quotient bit $Q_i = \bar{R}_0$.
Step 4. Shift partial remainder: $C \circ S = 2(C^i \circ S^i)$.
Step 5. If $i > n$, End. Otherwise, $i = i+1$, go to *Step 1*.

A numerical example is given in Figure 7.4 for illustration.

As a summary, the gate count and the delay time of the three types of array dividers are listed in the following table in terms of the quotient word length n, and the unit gate delay $\triangle_g$.

Type of Array Divider	Gate Count	Divide Time
Restoring	$14n^2 + 4n$	$(2n^2 + 6n + 2)\triangle_g$
Non-restoring	$15(n+1)^2$	$(2n^2 + 6n + 4)\triangle_g$
Non-restoring with CLA	$18n^2 + 27n + 10$	$(10n + 11)\triangle_g$

It should be pointed out that gate count is not the only concern in the area complexity consideration. A modern component may contain millions of gates, each taking up finite space. To work together these gates need to communicate with each other. Wires are required for interconnect, which sometimes may take up most of the area. As an example, Figure 7.5 shows a routed data encryption standard circuit in which the area associated with wires is significant[19].

REFERENCES

1. D. P. Agrawal, "High-Speed Arithmetic Arrays," *IEEE Trans. Comp.*, C-28 (Mar. 1979), pp. 215–224.

2. M. Cappa, "Cellular Iterative Arrays for Multiplication and Division," M.S. Thesis, Dept. of Elect. Eng., Univ. of Toronto, Canada, Oct. 1971.

3. M. Cappa and V. C. Hamacher, "An Augmented Iterative Array for High-Speed Binary Division," *IEEE Trans. Comp.*, Vol. C-22, Feb. 1973, pp. 172–175.

4. K. J. Dean, "Binary Division Using a Data Dependent Iterative Arrays," *Electronics Letters*, Vol. 4, July 1968, pp. 283–284.

5. J. Deverell, "The Design of Cellular Arrays for Arithmetic," *Radio and Electronic Engineer*, Vol. 44, No. 1, Jan. 1974, pp. 21–26.

6. D. Ferrari, "A Division Method Using a Parallel Multiplier," *IEEE Trans. Comp.*, Vol. EC-16, Apr. 1967, pp. 224–226.

7. M. J. Flynn, "On Division by Functional Iteration," *IEEE Trans. Comp.*, Vol. C-19, Aug. 1970, pp. 702–706.

8. A. B. Gardiner, Comments on "An Augmented Iterative Array for High-Speed Binary Division," *IEEE Trans. Comp.*, Vol. C-23, No. 3, Mar. 1974, pp. 326–327.

9. A. B. Gardiner and J. Hont, "Comparison of Restoring and Nonrestoring Cellular Array Dividers," *Electronics Letters*, Vol. 7, Apr. 1971, pp. 172–173.

10. A. Gex, "Multiplier-Divider Cellular Array," *Electronics Letters*, Vol. 7, July 1971, pp. 442–444.

11. H. H. Guild, "Some Cellular Logic Arrays for Non-restoring Binary Division," *Radio and Electronics Engineer*, Vol. 39, June 1970, pp. 345–348.

12. V. C. Hamacher and J. Gavilian, "High-Speed Multiplier/Divider Iterative Arrays," *Proc. 1973 Sagamore Computer Conf. on Parallel Processing*, 1973, pp. 91–100.

13. J. C. Majithia, "Nonrestoring Binary Division Using a Cellular Array," *Electronics Letters*, Vol. 6, May 1970, pp. 303–304.

14. J. E. Robertson, "Theory of Computer Arithmetic Employed in the Design of New Computer at the University of Illinois," Tech. Rept., No. 319, Dept. of Computer Science, University of Illinois, Urbana, 1960.

15. A. Socemcantu, "Cellular Logic Array for Redundant Binary Divisions," *Proc. IEE (London)*, Vol. 119, No. 10, Oct. 1972, pp. 1452–1456.

16. R. Stefanelli, "A Suggestion for High-Speed Parallel Binary Divider," *IEEE Trans. Comp.*, Vol. C-21, Jan. 1972, pp. 42–55.

17. A. Svoboda, "An Algorithm for Division," *Inf. Proc. Machines*, No. 9, Prague, Czechoslovakia, 1963, pp. 25–34.

18. M. V. Wilkes, D. J. Wheeler and S. Gill, *Preparation of Programs for an Electronic Digital Computer*, Addison-Wesley, Reading, MA, 1951.

19. A. Dehon, "Interconnect," Web page at http://www.cs.caltech.edu/cbsss/schedule/slides/dehon3_interconnect.pdf, June 20, 2002.

PROBLEMS

7.1 Given a dividend $A = (.1011011010)_2$ and a divisor $B = (.01011)_2$, implement the division through a 5-by-5 Restoring array divider. Draw the structure of the array and find the quotient and remainder following the data flow. What is the division time?

7.2 Construct a 3-by-3 restoring array divider to implement the division of $C = (.110101)_2$ by $D = (.101)_2$. Find the intermediate result after $6\Delta_g$ and after $9\Delta_g$ by showing the values of c_{out} and s for each cell.

7.3 Obtain $N = (.101101)_2$ divided by $D = (.111)_2$ through a non-restoring array divider. Show the S_{ij} and $C_ij + 1$ for each CAS cell (i,j), where $0 \le i \le n$ indicates the row index and $0 \le j \le n$ indicates the column index.

7.4 If the XOR and full adder in each cell of non-restoring array divider is realized by NAND-NAND implementation, assume all the input variables are available in both the original and the complement form. Find the division time of the array for an n-bit division.

7.5 Apply a carry-lookahead array divider to perform a 5-bit division with dividend $A = (.0001101110)_2$ and divisor $D = (.01010)_2$. Show the detailed procedure of this division with reference to Figure 7.3.

7.6 Given a 16-by-16 restoring divider, a 16-by-16 non-restoring array divider and a 16-by-16 non-restoring array divider with CLA, compare the cost effectiveness of them.
Hint: Find the area complexity and delay time of each array divider first.

8
Floating Point Operations

Floating point operations are widely applied in scientific computations. With limited number of digits, the range and precision of the numbers represented by floating point systems can be improved. In this chapter, we introduce floating point addition, subtraction, multiplication and division.

8.1 FLOATING POINT ADDITION/SUBTRACTION

Let $X_1 = (M_1, E_1)$ and $X_2 = (M_2, E_2)$ be two numbers in floating point representation, where $M_i = S_i |M_i|$ and $X_i = (-1)^{S_i} \cdot |M_i| \cdot r^{E_i - bias}$. We are to find $X_{out} = X_1 \pm X_2$.

Two floating point numbers cannot be added/subtracted unless the two exponents of them are equal. An *alignment* is needed if the exponents of the two given numbers are different. Usually, we let the bigger exponent remain unchanged, and adjust the smaller exponent to be the same as the bigger one. For a number with exponent enlarged, its mantissa should be reduced in order to keep the value of the number as same as before. That is, the mantissa should be shifted right. The exponent was increased by $|E_1 - E_2|$, resulting in $r^{|E_1 - E_2|}$ times enlargement. The number of digit positions to be right shifted in mantissa should be $|E_1 - E_2|$ as well resulting in $r^{|E_1 - E_2|}$ times reduction (indicated by a factor of $r^{-(|E_1 - E_2|)}$ below).

Let $X_{out} = X_1 \pm X_2 = (M_{out}, E_{out})$. We have

$$E_{out} = max\{E_1, E_2\},$$

183

and

$$M_{out} = \begin{cases} M_1 \pm M_2 \cdot r^{-|E_1-E_2|}, & \text{if } E_1 > E_2 \\ M_1 \cdot r^{-|E_1-E_2|} \pm M_2, & \text{if } E_1 \leq E_2. \end{cases}$$

The addition/subtraction procedure includes the following steps.

1. *Alignment.* Shift the mantissa of the smaller operand to the right. The number of digit positions to shift over is equal to the difference between the two exponents. The larger exponent is the exponent of the result.

2. *Mantissa Addition/Subtraction.* One mantissa is added to or subtracted from the other.

3. *Postnormalization.* Normalize the mantissa resulted in sum/difference if necessary, and adjust the exponent accordingly.

For example, suppose $r = 2$,
$X_1 = (M_1, E_1) = (1001, 10)$, and
$X_2 = (M_2, E_2) = (1100, 00)$.

Find $X_{out} = X_1 - X_2$.

$$X_{out} = (1001, 10) - (1100, 00)$$

$$\Downarrow \text{alignment}$$

$$X_{out} = (1001, 10) - (0011, 10)$$

$$\Downarrow \text{subtraction}$$

$$X_{out} = (0110, 10)$$

$$\Downarrow \text{postnormalization}$$

$$X_{out} = (1100, 01)$$

The data flow of the floating point subtraction/addition is shown in Figure 8.1. OMZ refers to order-of-magnitude zero, particularly the scenario $(M, E) = (0, E)$ where $E \neq 0$. It can be resulted from the mantissa subtraction when the two mantissas are equal, and is in contrast to true zero, in which $(M,E) = (0,0)$. Since it is impossible to perform a postnormalization over a mantissa equal to zero, OMZ should be detected and its existence signaled.

8.2 FLOATING POINT MULTIPLICATION

In a floating point multiplication, the two mantissas are to be multiplied, and the two exponents are to be added. That is, if $X_1 = M_1 \cdot r^{E_{1\,unbiased}}$ and $X_2 = M_2 \cdot$

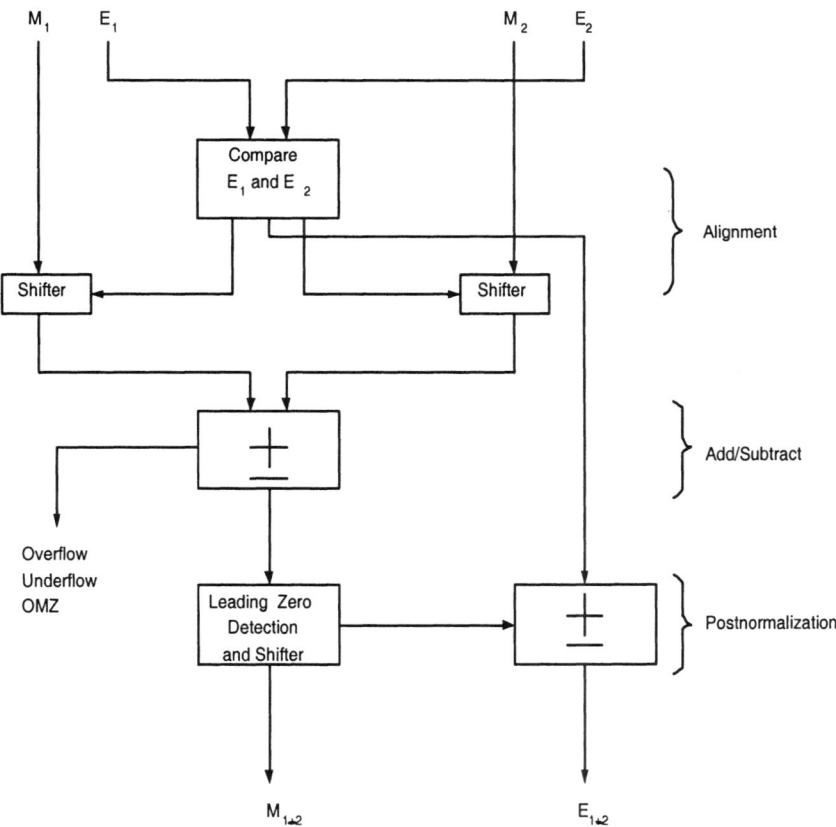

Fig. 8.1: Data Flow of Floating Point Addition/Subtraction

$r^{E_{2\,unbiased}}$, then

$$X_p = X_1 \times X_2 = M_p \cdot r^{E_{p\,unbiased}},$$

where

$$M_p = M_1 \times M_2,$$

and

$$E_{p\,unbiased} = E_{1\,unbiased} + E_{2\,unbiased}.$$

Given the normalized mantissas M_1 and M_2, the product of them may not be normalized, and a postnormalization is necessary in that case. A left shift for one and at most one digit position will be needed for postnormalization. Consider the binary representation, for example. $(0.1 \times \cdots \times) \times (0.1 \times \cdots \times)$ will result in $(0.01 \times \cdots \times)$. In general, $\frac{1}{r} \leq M_1 < 1$ and $\frac{1}{r} \leq M_2 < 1$ will result in $\frac{1}{r^2} \leq M_1 \times M_2 < 1$.

With biased exponents, we can represent X_1 and X_2 by $X_1 = (M_1, E_1)$ and $X_2 = (M_2, E_2)$, where

$$E_1 = E_{1\,unbiased} + bias \tag{8.1}$$

and

$$E_2 = E_{2\,unbiased} + bias. \tag{8.2}$$

The product is to be represented by $X_p = (M_p, E_p)$, where E_p is a biased exponent and

$$\begin{aligned} E_p &= E_{p\,unbiased} + bias \tag{8.3} \\ &= E_{1\,unbiased} + E_{2\,unbiased} + bias. \tag{8.4} \end{aligned}$$

Adding Equation (8.1) and Equation (8.2), and comparing the sum with the right-hand side of Equation (8.4), we can find that the latter is one bias less. So, E_p can be found by

$$E_p = E_1 + E_2 - bias.$$

The subtraction is necessary because otherwise, the bias is included twice.

The data flow of the floating point multiplication is shown in Figure 8.2. Following is an example for binary floating point multiplication.

Suppose $r = 2$, $X_1 = (1010, 01)$, and $X_2 = (1010, 10)$. Find $X_p = X_1 \times X_2$. Here, bias $= 2^1 - 1 = 1$.

$$\begin{aligned} X_p &= (1010 \times 1010, (01 + 10 - 1)) \\ &= (01100110, 10) \\ &= (1100, 01). \end{aligned}$$

$$X_1 \times X_2 = (M_1, E_1) \times (M_2, E_2) = (M_p, E_p)$$

Data Flow :

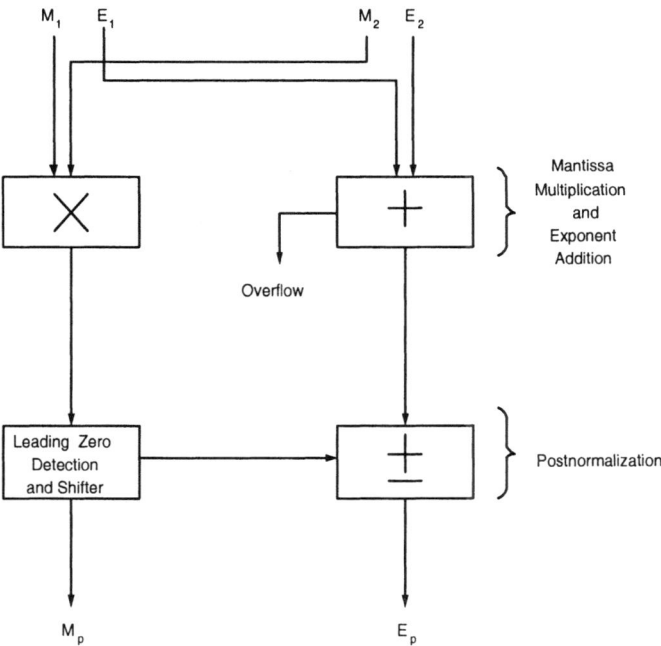

Fig. 8.2: Data Flow of Floating Point Multiplication

8.3 FLOATING POINT DIVISION

In a floating point division, the mantissa of dividend is divided by that of divisor, and the exponent of dividend minus that of divisor will be resulted. That is, if $X_1 = M_1 \cdot r^{E_{1 unbiased}}$ and $X_2 = M_2 \cdot r^{E_{2 unbiased}}$, then

$$X_p = X_1/X_2 = M_p \cdot r^{E_{p unbiased}},$$

where

$$M_p = M_1/M_2,$$

and

$$E_{p unbiased} = E_{1 unbiased} - E_{2 unbiased}.$$

Given the normalized mantissas M_1 and M_2, we have $\frac{1}{r} \leq M_1 < 1$ and $\frac{1}{r} \leq M_2 < 1$. Referring $\frac{1}{r}$ as M_{min} and 1 as M_{max}, the range of M_1/M_2 is $M_{min}/M_{max} < M_1/M_2 < M_{max}/M_{min}$, that is,

$$\frac{1}{r} < M_1/M_2 < r.$$

The left side of the inequality indicates that the quotient is always normalized and no postnormalization is needed. The right side of the inequality shows a possibility of quotient overflow when $1 < M_1/M_2$. A right shift over one digit position will solve the problem and reduce any quotient from $< r$ to < 1.

Given X_1 and X_2 represented with biased exponents such as $X_1 = (M_1, E_1)$ and $X_2 = (M_2, E_2)$ for the same E_1 and E_2 defined in Equations (8.1) and (8.2), the quotient can be represented by $X_p = (M_p, E_p)$ where E_p is a biased exponent and

$$\begin{align} E_p &= E_{p unbiased} + bias & (8.5)\\ &= E_{1 unbiased} - E_{2 unbiased} + bias. & (8.6) \end{align}$$

Subtract Equation (8.2) from Equation (8.1), the two biases on the right-hand side cancel, and we have

$$E_1 - E_2 = E_{1 unbiased} - E_{2 unbiased}.$$

Comparing with Equation (8.3), we can see that E_p can be found from the difference of E_1 and E_2 if a bias is added back. That is,

$$E_p = E_1 - E_2 + bias.$$

The data flow of the floating point division is shown in Figure 8.3. Following is an example for binary floating point division.

Suppose $r = 2$,
$X_1 = (1010, 01)$, and

$$X_1 / X_2 = (M_1, E_1) / (M_2, E_2) = (M_q, E_q)$$

Data Flow :

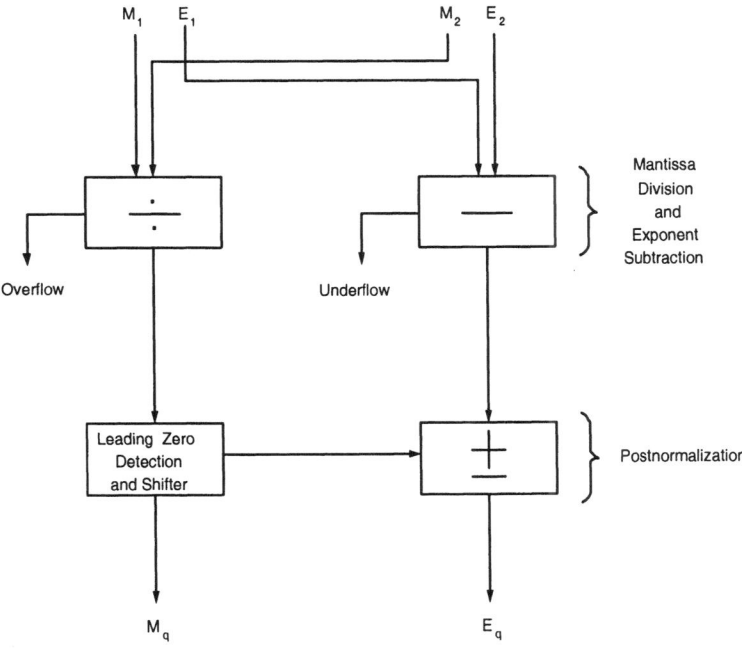

Fig. 8.3: Data Flow of Floating Point Division

$X_2 = (1010, 00)$.
Find $X_p = X_1/X_2$. Notice that $bias = 1$.

$$\begin{aligned} X_p &= (1010/1010, (01 - 00 + 1)) \\ &= (1\ 0000, 10) \\ &= (1000, 11). \end{aligned}$$

8.4 ROUNDING

Precisely representing a number with limited digits is always an issue in computer systems. When only part of the digits in the result of calculation can be retained, the rounding problem arises. Rounding is important particularly in floating point arithmetic since usually a sequence of computations are performed for scientific work, while the error caused by rounding accumulates and the error bound increases more rapidly.

In the example in Section 8.2, one can see that for a resulted product with $2n$ digits, we retain only n digits. With limited number of digits, it is often hard to represent the exact value of a result, say X_o. In digit systems, however, one can always find X_a and X_b that can be represented and can have X_o tightly bounded. In other words, $X_a \leq X_o \leq X_b$, where X_a is the greatest of those representable smaller than X_o, and X_b the smallest of those representable greater than X_o.

The rounding policies include truncate, round-up and round-to-nearest which are explained as follows. We first focus on positive numbers for simplicity.

1. *Truncate*: The lower order digits are discarded which are less significant.

2. *Round-up*: The lower order digits are discarded and, at the mean time, the least significant digit of the higher order digits which are retained is incremented.

3. *Round-to-nearest*: If p digits are to be retained, the $(p+1)$th digits is to be examined. If it is $< \frac{r}{2}$, discard it and every digit on its right. If it is $\geq \frac{r}{2}$, discard it as well as every digit on its right, and add 1 to the pth digit. Round-to-nearest is a combination of the truncate and round-up approaches mentioned above, executing one of them based on the case condition rather than all the time.

In decimal number system, for example, suppose only integer digits are to be kept. If truncating is adopted, 2.4 will become 2, and 2.5 will become 2 as well. If round-up is adopted, 2.4 will become 3 and 2.5 will become 3. If round-to-nearest is adopted, 2.4 will become 2, and 2.5 will become 3.

Consider the rounding of 2.5000 with round-to-nearest approach. The difference of 2.5000 and 2 is 0.5, and the difference of 2.5000 and 3 is also 0.5. That is, 2.5000 is within equal distance to integer 2 and 3 (referred to as in half-way). It is rounded to 3 as defined by round-to-nearest approach, though it is not nearer to 3 than to 2. It is not fair to always round it to 3, and it is not fair neither to always round it to 2. A better way to reduce the accumulative error is, as indicated in IEEE rounding standard, to "round to nearest even". For the above example, that is, round 2.5000 to 2. Between the two integers serving as the lower bound and upper bound of $x.5$ for any integer x, one must be even and one odd. Round to that even number, according to the "round to nearest even" approach, no matter whether truncate is needed or round-up is needed.

Notice that in the above example, all the digits following .5 should be examined to assure that half-way case is dealt with. If any digit on the right of .5 is more than 0, we don't have a half-way case; Round-to-nearest approach in (3) can be implemented, since it is easy to find which integer is nearer.

In Table 8.1, the given 6-bit numbers are to be rounded to 4-bit ones, and "round to nearest even" policy is to be adopted.

Take negative numbers into consideration. If the negative numbers are represented in sign-magnitude form, and the above listed rounding procedure works on the magnitudes of the numbers, then, truncating a magnitude and deleting the digits on its less

Table 8.1: Round to Nearest Even

1001	00 → 1001	truncate
1001	01 → 1001	truncate
1001	10 → 1010	mid-way of 1001 and 1010 round to 1010
1001	11 → 1010	round-up
1010	00 → 1010	truncate
1010	01 → 1010	truncate
1010	10 → 1010	mid-way of 1010 and 1011 round to 1010
1010	11 → 1011	round-up

significant end make the magnitude smaller, and lead the signed number "round-to-zero". (For negative numbers represented in complete form, truncating them rounds away from zero instead of toward zero.) When round-up a magnitude, for a positive number, it is rounded toward $+\infty$, and for a negative number it is rounded toward $-\infty$.

8.5 EXTRA BITS

Let's compare two examples,
1. $X_{out} = (1001, 10) - (1010, 00)$, and
2. $X'_{out} = (1001, 10) - (1011, 00)$.

If only 4-bit subtraction is allowed, we have

$X_{out} = (1001, 10) - (1010, 00)$
$X'_{out} = (1001, 10) - (1011, 00)$

$\Downarrow$ alignment

$X_{out} = (1001, 10) - (0010, 10) = X'_{out}$

$\Downarrow$ subtraction

$X_{out} = (0111, 10) = X'_{out}$

$\Downarrow$ postnormalization

$X_{out} = (1110, 01) = X'_{out}$

The two subtractions ended up with the same result! This is because that in the alignment step, some bits were shifted to right and shifted out, and part of the information about the subtrahends was lost. In the postnormalization, both of the differences were shifted left and shifted back for one bit position, but both shifted in a 0.

If a 6-bit subtraction is allowed, after the alignment, the subtrahend in Example 1 becomes 001010, and that in Example 2 becomes 001011. We have

```
        Example 1              Example 2

        100100                 100100
    -)  001010             -)  001011
        011010                 011001
```

A postnormalization is needed here. After left shift the mantissa for one bit position, the two results become

$$X_{out} = (1101, 10),$$

and

$$X'_{out} = (1100, 10).$$

Here, a difference of 1 can be observed in the LSB of mantissas. That bit was the fifth bit in the subtraction, and in both of the examples the fifth bit was retained, rather than shifted out. The implication we obtain here is to retain some extra bits in the operation for the sake of accuracy.

How many extra bits to keep is worth investigation, then. For multiplication, we understand from Section 8.2 that the product of two normalized mantissas has at most one leading zero. In the postnormalization, left shift is required and one bit is to be shifted in from right.

For addition/subtraction, if a subtrahend has been shifted to right for two or more bit positions in an earlier alignment step, then at most one leading 0 will be caused after the subtraction. This is because that given a normalized minuend, the smallest value of it is

$$0.10 \cdots 0;$$

and if a normalized subtrahend was shifted right for two or more bit positions for the alignment, the maximal value of it is

$$0.001 \cdots 1.$$

Hence

```
         0.100 ··· 00
    -)   0.001 ··· 11
```

$$0.010\cdots 01$$

is the worst case. The result is to be shifted left for postnormalization, and one bit is to be shifted in from right. If one extra bit is kept on the right, its true value can be shifted in, rather than a 0 to be shifted in all the time. Such extra bit is referred to as *guard bit*.

Another extra bit can be added to the right of the guard bit for rounding function, referred to as *round bit*. Instead of truncating all the bits in less significant bit positions all the time, we conduct the truncation only if the round bit is 0. If the round bit is 1, we round-up. In other words, with the round bit, round-to-nearest can be realized. For an even fairer process "round to nearest even", the half-way case should be identified.

In the above case, when the round bit is 1, we round up, assuming that the amount represented by it and the lower order bits is more than half of the weight carried by LSB. However, when the round bit is 1 and all the bits on its right are 0's, a half-way case presents. Half-way cases are not always rounded up unless an even number can be resulted after rounding. Note, to be a half-way case, no bit on the right of round bit is allowed to be 1, including those bits shifted out in the alignment step. The information of those bits should not be lost, and to record it another extra bit, called *sticky bit*, is added to the right of round bit. Every bit passing sticky bit position on its way of shifting right will be examined. If any bit is 1, the sticky bit sticks to 1. Actually, the sticky bit (S in below) is the logic OR of all the bits passing through, providing a summarized information for all the bits on the right of round bit (R in below).

$$\times \quad 1 \quad \underbrace{0 \cdots 0}$$
$$G \quad R \quad S$$

Truncating R and the lower order bits, we have the lower bound of the given number. When LSB = 0, the lower bound is even. When LSB = 1, the lower bound is odd and the upper bound is to be used since it must be even. The upper bound can be obtained by adding 1 to the LSB bit position. Denote LSB as B_0, the following logic expression describes when the add 1 function should be performed.

$$R \cdot S + B_0 \cdot R \cdot \bar{S} \tag{8.7}$$
$$= R(S + \bar{S} \cdot B_0) \tag{8.8}$$
$$= R(S + B_0) \tag{8.9}$$
$$= R \cdot S + R \cdot B_0 \tag{8.10}$$

The second term is for the half-way case ($R = 1$ and $S = 0$) and when the given number is odd ($B_0 = 1$) in which we round up. The first term is for the "more than half" case ($R = 1$ and $S = 1$) in which we round up no matter the given number is odd or even. Equation (8.7) has actually formulated the IEEE rounding policy: round to the nearest, and round to even for the half-way case. Figure 8.4 shows an example of rounding in a subtraction.

194 FLOATING POINT OPERATIONS

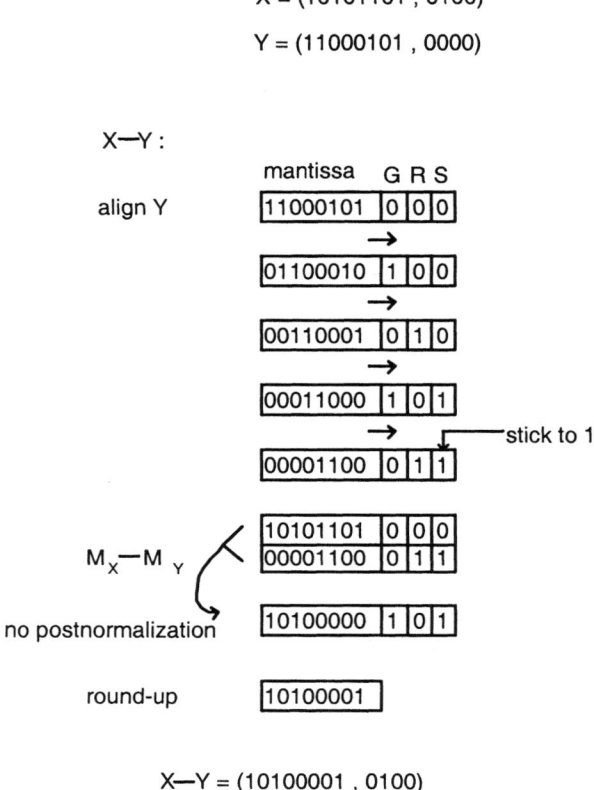

Fig. 8.4: Example of Rounding in Subtraction

the nearest, and round to even for the half-way case. Figure 8.4 shows an example of rounding in a subtraction.

REFERENCES

1. G. M. Amdahl, "The Structure of System/360, Part III, Processing Unit Considerations," *IBM System Journal*, Vol. 3, 1964, pp. 144–164.

2. S. F. Anderson, et. al., "The IBM System/360 Model 91: Floating-Point Execution Unit," *IBM Journal*, Jan. 1967, pp. 34–53.

3. R. P. Brent, "On the Precision Attainable with Various Floating-Point Number Systems," *IEEE Trans. Comp.*, C-22 (June 1973), pp. 601–607.

4. S. G. Campbell, *Floating-Point Operation, Planning a Computer System*, W. Buckholz ed., McGraw-Hill, New York, 1962, Chapter 8.

5. W. J. Cody, Jr., "Analysis of Proposals for the Floating-Point Standard," *Computer*, (Mar. 1981), pp. 63–69.

6. W. J. Cody, Jr., "Static and Dynamic Numerical Characteristics of Floating-Point Arithmetic," *IEEE Trans. Comp.*, Vol. C-28, June 1973, pp. 596–601.

7. Y. Chu, *Computer Organization and Microprogramming*, Prentice-Hall, Englwood Cliffs, NJ, 1972, Chapter 5.

8. M. J. Flynn and P. R. Low, "The IBM System/360 Model 91: Some Remarks on System Development," *IBM Journal*, Jan. 1967, pp. 2–7.

9. H. L. Gary, and C. Harrison, Jr., "Normalized Floating-Point Arithmetic with an Index of Significance," *Proc. Eastern Joint Computer Conference*, 1959, pp. 244–248.

10. D. Goldberg, "Computer Arithmetic," *in Computer Architecture: A Quantitative Approach*, D. A. Patterson and J. L. Hennessy, Morgan Kaufmann, San Mateo, CA, 1990, App. A.

11. IEEE Standard for Binary Floating-Point Arithmetic, ANSI/IEEE 754-1985, also in *Computer*, 14 (Mar. 1981), pp. 51–62.

12. IEEE Task P 754, "A Proposed Standard for Binary Floating-Point Arithmetic," *IEEE Comp.*, Vol. 14, No. 3, Mar. 1981, pp. 51–62.

13. IBM Staff, "Floating-Point Arithmetic," in IBM System/360 Principles of Operation, IBM System Ref. Lib. From A22-6821-7, September 1968, pp. 41–50.3.

14. W. Kahan, "What is the Best Base for Floating-Point Arithmetic, Is Binary Best?" Lecture Notes, Dept. of Computer Sci., University of California, Berkeley, 1970.

15. U. Kulisch, "Mathematical Foundations of Computer Arithmetic," *IEEE Trans. Comp.*, Vol. C-26, No. 7, July 1977, pp. 610–621.

16. D. E. Knuth, *The Art of Computer Programming: Seminumerical Algorithms*, Vol. 2, Addison-Wesley, Reading, 1969, Chapter 4.

17. D. J. Kuck, D. S. Parker, and A. H. Sameh, "Analysis of Rounding Methods in Floating- Point Arithmetic," *IEEE Trans. Comp.*, Vol. C-26, No. 7, July 1977, pp. 643–650.

18. L. A. Liddiard, "Required Scientific Floating Point Arithmetic," *Proc. 4th Symp. on Comp. Arith.*, Oct. 1978, IEEE Cat. No. 78CH1412-6C, pp. 56–62.

19. D. W. Matula, "A Formalization of Floating-Point Numeric Base Conversion," *IEEE Trans. Comp.*, Vol. C-19, August 1970, pp. 681–692.

196 FLOATING POINT OPERATIONS

20. M. M. Mano, *Computer System Architecture*, Prentice-Hall, Englewood Cliffs, NJ, 1976, Chapter 10.

21. R. E. Moore, *Interval Analyis*, Prentice-Hall, Englewood Cliffs, NJ, 1966.

22. E. K. Reuter, et al., "Some Experiments Using Interval Arithmetic," *Proc. 4th Symp. on Comp. Arith.*, Oct. 1978, IEEE Cat. No. 78CH1412-6c, pp. 75–80.

23. P. L. Richman, "Floating-Point Number Representations: Base Choice Verus Exponent Range," Tech. Rep. No. CS-64, Dept. of Comp. Sci., Stanford University, Stanford, CA, 1967.

24. M. R. Santoro, G. Bewick, and M. A. Horowitz, "Rounding Algorithms for IEEE Multipliers," *Proc. 9th Symp. on Comp. Arith.*, 1989, pp. 176–183.

25. D. W. Sweeney, "An Analysis of Floating-Point Addition," *IBM Systems Journal*, Vol. 4, No. 1, 1965, pp. 31–42.

26. P. H. Sterbenz, *Floating-Point Computation*, Prentice-Hall, Englewood Cliffs, NJ, 1974.

27. R. M. Tomasulo, "An Efficient Algorithm for Exploiting Multiple Arithmetic Units," *IBM Journal*, Jan. 1967, pp. 25–33.

28. N. Tsao, "On the Distribution of Significant Digits and Roundoff Errors," *Communications of the ACM*, 17 (May 1974), pp. 269–271.

29. J. H. Wilkinson, *Rounding Errors in Algebraic Processes*, Prentice-Hall, Englewood Cliffs, NJ, 1963.

30. J. M. Yohe, "Roundings in Floating-Point Arithmetic," *IEEE Comp.*, Vol. C-22, No. June 1973, pp. 577–586.

PROBLEMS

8.1 IBM 370 system has a floating point format (short) similar to what is shown in Figure 1.8, except that E is of 7 bits and $|M|$ 24 bits. The bias is 64, and the normalized mantissa contains fraction only. The base r is chosen to be 16. Represent 17.5×2^{10} in such a floating point form. Write the mantissa and exponent in hexadecimal form.

8.2 What is the decimal value of the following number represented in the IEEE single-precision floating point form?

1	1000 0001	10000000000000000000000

8.3 To represent a number in floating point form, if 16 bits are used with 5 bits for the exponent which is unbiased, and 10 bits for the mantissa which contains normalized fraction only, what are the range and precision (in decimal) of the number that can be represented if base r is chosen to be 2. Justify your answer.

8.4 Given 32 bits to represent a number in fixed-point arithmetic or in floating point arithmetic with IEEE standard form, compare the range of the number that can be represented and the precision in number representation.

8.5 Given each operand represented in the basic binary floating point form, perform the following operations. Show step by step the arithmetic procedure and the contents of relevant registers.
(a) $0.625 \times 2^5 + 0.75 \times 2^{10}$
(b) $0.125 \times 2^{15} - 0.625 \times 2^8$
(c) $-0.875 \times 2^{12} + 0.25 \times 2^7$
(d) $-0.375 \times 2^6 - 0.5 \times 2^4$

8.6 Perform the following operations showing step by step the arithmetic procedure and the contents of relevant registers. Suppose each operand is represented in the basic binary floating point form.
(a) $0.125 \times 2^{10} \times 0.25 \times 2^8$
(b) $0.75 \times 2^{15} \times 0.625 \times 2^5$
(c) $0.375 \times 2^6 \div 0.5 \times 2^7$
(d) $0.875 \times 2^{12} \div 0.125 \times 2^4$

8.7 How many guard bits are needed to assure the accuracy in the results of subtraction? Justify your solution.

8.8 Following are the mantissas of the results obtained by floating point computations. Assuming that only one integer bit and four fraction bits can be retained, round the mantissas according to the IEEE "round to nearest even" policy after post normalization.

			.	1	0	1	0	1	0
								r	s
	1	0	.	1	1	1	0	0	1
								r	s
	1	0	.	0	0	1	1	0	0
								r	s
	1	1	.	1	1	0	1	0	1
								r	s
	1	1	.	0	0	1	1	0	1
								r	s

8.9 If someday IEEE changes the rounding policy to "round to nearest odd," write a logic expression to implement such rounding policy through "add 1" operation to the LSB position.

9
Residue Number Operations

Since in Residue Number Systems (RNS) the moduli are independent of each other, there is no carry propagation among them. The operations based on each modulus can be performed in parallel, and the RNS computations can be completed more quickly – an attractive feature for people who need high-speed arithmetic operations.

9.1 RNS ADDITION, SUBTRACTION AND MULTIPLICATION

Suppose two numbers, X and Y, are represented as
$$X = (x_1, x_2, \cdots, x_n)$$
and
$$Y = (y_1, y_2, \cdots, y_n)$$
in RNS. We use $\otimes$ to represent the operator of addition, subtraction and multiplication, and let
$$Z = X \otimes Y.$$
If
$$Z = (z_1, z_2, \cdots, z_n),$$
the arithmetic in RNS can be expressed as follows:
$$z_i = |x_i \otimes y_i|_{m_i}.$$

For example, if $m_1 = 5$, $m_2 = 3$, $m_3 = 2$, then

$$\begin{array}{rl} 4 & (4,1,0) \\ +\ 8 & +\ (3,2,0) \\ \hline 12 & (2,0,0) \end{array} \Longrightarrow$$

and

$$\begin{array}{rl} 3 & (3,0,1) \\ \times\ 7 & \times (2,1,1) \\ \hline 21 & (1,0,1). \end{array} \Longrightarrow$$

From the definition of the mod operation, all moduli are positive. x_i may be less than y_i which yields $x_i - y_i < 0$. In the mod operation, if $x_i - y_i < 0$, then z_i is defined as

$$z_i = m_i + (x_i - y_i). \tag{9.1}$$

Overflow detection, sign detection, number comparison and division in RNS are very difficult and time consuming. These shortcomings limit most of the previous RNS applications to additions, subtractions and multiplications.

9.2 NUMBER COMPARISON AND OVERFLOW DETECTION

With the representation in residues, number comparison and overflow detection in RNS have never been easy tasks. In this section, efficient methods are sought for number comparison and overflow detection. One can see that the methods introduced here are practical and can be easily implemented.

9.2.1 Unsigned Number Comparison

Let parity indicate whether an integer number is even or odd. We say that two numbers are of the same parity if they are both even or both odd. Otherwise the two numbers are said to be of different parities. We will apply the properties of the parities of numbers to accomplish the number comparison.

Let X and Y have the same parity and $Z = X - Y$. $X \geq Y$, if and only if (*iff*) Z is an even number. $X < Y$, *iff* Z is an odd number.

Actually, if $X \geq Y$, then $X - Y \geq 0$ and Z equals $X - Y$. We know from the mathematical axioms that if the two numbers are with the same parity, the result of subtraction should be an even number. Therefore $X \geq Y$ implies that Z is an even number.

On the other hand, suppose Z is an even number and X and Y are with the same parity. If $X < Y$ then $X - Y < 0$. From equation (1) we have $Z = X - Y + M$. Since M is an odd number and $X - Y$ is even, Z must be an odd number. This contradicts the assumption that Z is even. Therefore if Z is an even number and X and Y are with the same parity, then $X \geq Y$.

Table 9.1: Parity Table for Modulus Set $\{3, 5, 7\}$

#	(3 5 7)	P	#	(3 5 7)	P	#	(3 5 7)	P	#	(3 5 7)	P	#	(3 5 7)	P
0	0 0 0	0	21	0 1 0	1	42	0 2 0	0	63	0 3 0	1	84	0 4 0	0
1	1 1 1	1	22	1 2 1	0	43	1 3 1	1	64	1 4 1	0	85	1 0 1	1
2	2 2 2	0	23	2 3 2	1	44	2 4 2	0	65	2 0 2	1	86	2 1 2	0
3	0 3 3	1	24	0 4 3	0	45	0 0 3	1	66	0 1 3	0	87	0 2 3	1
4	1 4 4	0	25	1 0 4	1	46	1 1 4	0	67	1 2 4	1	88	1 3 4	0
5	2 0 5	1	26	2 1 5	0	47	2 2 5	1	68	2 3 5	0	89	2 4 5	1
6	0 1 6	0	27	0 2 6	1	48	0 3 6	0	69	0 4 6	1	90	0 0 6	0
7	1 2 0	1	28	1 3 0	0	49	1 4 0	1	70	1 0 0	0	91	1 1 0	1
8	2 3 1	0	29	2 4 1	1	50	2 0 1	0	71	2 1 1	1	92	2 2 1	0
9	0 4 2	1	30	0 0 2	0	51	0 1 2	1	72	0 2 2	0	93	0 3 2	1
10	1 0 3	0	31	1 1 3	1	52	1 2 3	0	73	1 3 3	1	94	1 4 3	0
11	2 1 4	1	32	2 2 4	0	53	2 3 4	1	74	2 4 4	0	95	2 0 4	1
12	0 2 5	0	33	0 3 5	1	54	0 4 5	0	75	0 0 5	1	96	0 1 5	0
13	1 3 6	1	34	1 4 6	0	55	1 0 6	1	76	1 1 6	0	97	1 2 6	1
14	2 4 0	0	35	2 0 0	1	56	2 1 0	0	77	2 2 0	1	98	2 3 0	0
15	0 0 1	1	36	0 1 1	0	57	0 2 1	1	78	0 3 1	0	99	0 4 1	1
16	1 1 2	0	37	1 2 2	1	58	1 3 2	0	79	1 4 2	1	100	1 0 2	0
17	2 2 3	1	38	2 3 3	0	59	2 4 3	1	80	2 0 3	0	101	2 1 3	1
18	0 3 4	0	39	0 4 4	1	60	0 0 4	0	81	0 1 4	1	102	0 2 4	0
19	1 4 5	1	40	1 0 5	0	61	1 1 5	1	82	1 2 5	0	103	1 3 5	1
20	2 0 6	0	41	2 1 6	1	62	2 2 6	0	83	2 3 6	1	104	2 4 6	0

The above property shows us a method to compare two numbers if the parities of these two numbers are the same. Similarly if the parities of two numbers are different, then the following property can tell us which one is bigger.

Let X and Y have different parities and $Z = X - Y$. $X \geq Y$, iff Z is an odd number. $X < Y$, iff Z is an even number.

To prove the above we can see that if $X \geq Y$, then $X - Y \geq 0$ and Z equals $X - Y$. We know that the two numbers are with different parities, and the result of the subtraction should be an odd number. Therefore $X \geq Y$ implies that Z is an odd number.

On the other hand, suppose that Z is an odd number and X and Y are with different parities. If $X < Y$, then $X - Y < 0$. From Equation (9.1) we have $Z = X - Y + M$. Since M is an odd number and $X - Y$ is odd, Z must be an even number. This contradicts the assumption that Z is odd. Therefore if Z is an odd number and X and Y are with different parities, then $X \geq Y$.

Table 9.1 is referred to when performing parity checking for number comparisons. The decimal numbers under the entry "#" are corresponding to the residue numbers for modulus set $\{3, 5, 7\}$, and the parities of them are given under the entry P.

Following is an example illustrating the above theory.

Let the moduli be $m_1 = 3$, $m_2 = 5$, $m_3 = 7$ and hence $M = 3 \times 5 \times 7 = 105$. Consider $X_1 = (0, 3, 5)$ and $Y_1 = (1, 3, 0)$. From calculation we have $Z_1 = X_1 - Y_1 = (2, 0, 5)$. By table look-up, the parities of X_1, Y_1 and Z_1 are odd, even and odd respectively. From the above theory we know $X_1 > Y_1$.

In the decimal number system, $X_1 = 33$, $Y_1 = 28$ and $Z_1 = 5$, and the result is obvious.

9.2.2 Overflow Detection

Let m_is be all odd and pair-wise relatively prime. If two numbers are with the same parity, then the overflow in the addition of the two numbers can be detected as follows.

Suppose X and Y have the same parity and $Z = X + Y$. The addition is with overflow, *iff* Z is an odd number.

This is due to the fact that if X and Y have the same parity, $(X + Y)$ should result in an even number. From the RNS operation, if $(X + Y) \geq M$ then $Z = |X + Y|_M = X + Y - M$. However M is an odd number, which causes Z to be odd.

On the other hand, suppose Z is an odd number and X and Y are with the same parity. If $(X+Y)$ is not overflowed, then $(X+Y)$ is an even number. This contradicts the assumption that Z is an odd number. Therefore if Z is an odd number, and X and Y are with the same parity, then $(X + Y)$ is with overflow.

If two numbers are with different parities, then the overflow detection in the addition of the two numbers can be detected as follows.

Suppose that X and Y have different parities and $Z = (X + Y)$. The addition is with overflow, *iff* Z is an even number. The following example helps to describe the above theory.

Let the moduli be $m_1 = 3$, $m_2 = 5$, $m_3 = 7$, and two numbers be $X_1 = (2, 1, 1)$ and $Y_1 = (0, 0, 4)$. Detect whether the addition of X_1 and Y_1 is with overflow. From calculation $Z_1 = X_1 + Y_1 = (2, 1, 5)$. By looking up Table 9.1, the parities of X_1, Y_1 and Z_1 are found to be odd, even and even respectively. From the above theory, the result of the addition is overflowed. Actually $X_1 = 71$, $Y_1 = 60$ and $Z_1 = |131|_{105} = 26$ in the decimal number system.

The overflow detection described above applies to the addition of only two numbers.

9.2.3 Signed Numbers and Their Properties

The method to represent negative numbers in RNS is similar to that in conventional radix number systems. Letting the dynamic range be M, we can define the positive and negative numbers as follows.

Given m_is in the modulus set are all odd, and the dynamic range $M = \prod_{i=1}^{n} m_i$, then the range of a positive number X is defined as $0 \leq X \leq \lfloor \frac{M}{2} \rfloor$ and the range of a negative number Y is defined as $\lfloor \frac{M}{2} \rfloor < Y < M$. For any positive number $X \neq 0$, the additive inverse of X is represented by $(M - X)$. Notice here 0 is considered a positive number. The complement of X is $(M - X)$. In a similar way, the representation of the complement of a number in RNS can be found as follows.

Let the modulus set be $\{m_1, m_2, \cdots, m_n\}$, and the corresponding modulus set of a positive number X in RNS be $\{x_1, x_2, \cdots, x_n\}$. $-X$ in RNS can be represented by the complement of X which is equal to $(|m_1 - x_1|_{m_1}, |m_2 - x_2|_{m_2}, \cdots, |m_n - x_n|_{m_n})$. From the definition, $-X$ in RNS corresponds to $M - X$. Applying Equation (9.1), the corresponding modulus set of $-X$ is $(|m_1 - x_1|_{m_1}, |m_2 - x_2|_{m_2}, \cdots, |m_n - x_n|_{m_n})$.

The dynamic range of RNS can be divided into two halves, one for positive numbers and the other one for negative numbers. If the moduli are all pair-wise relatively primed and all odd numbers, then the maximum positive number is $\frac{M-1}{2}$. A negative number's magnitude must fall in the positive range. In this case, the unsigned number comparisons are applicable for the signed RNS numbers. When comparing two signed numbers, three cases should be considered. One number is positive and one negative, two numbers are both positive, and two numbers are both negative.

1. If X and Y are with different signs, the positive number is greater than the negative number.

2. If X and Y are both positive numbers, the unsigned number comparison in 9.3.2 can be applied to compare X and Y.

3. If X and Y are both negative numbers, then find the absolute values for X and Y and compare them. The number with a greater absolute value is smaller.

To define the overflows in signed RNS numbers, we have the following.

Given m_is in the modulus set all odd and two numbers in RNS such as $X = (x_1, x_2, ..., x_n)$ and $Y = (y_1, y_2, \cdots, y_n)$. Overflow exits if $|X + Y| > \frac{M-1}{2}$.

Note that the following cases should be considered. (1) X and Y are with the same sign. The absolute value of the sum should be no greater than $\lfloor \frac{M-1}{2} \rfloor$. (2) X and Y have different signs, no overflow will occur.

The above overflow detection theory applies to the addition of only two numbers.

9.2.4 Multiplicative Inverse and the Parity Table

Consider the number $|b|_m$. The multiplicative inverse of $|b|_m$ is defined as follows.

If $0 \leq a < m$ and $|ab|_m = 1$, a is called the multiplicative inverse of ($b \bmod m$) and is denoted as $|b^{-1}|_m$ or $|\frac{1}{b}|_m$.

Notice that the multiplicative inverse of a number does not always exist. The condition of its existence is described below.

The quantity $|b^{-1}|_m$ exists if and only if the greatest common divisor of b and m, $gcd(b, m)$, is equal to 1, and $|b|_m \neq 0$. In this case $|b^{-1}|_m$ is unique.

We used the parity checking technique to compare numbers and detect overflows in the addition of two numbers. For the parity checking, a redundant modulus 2 is required. The parity of a number, 0 if it is even or 1 if odd, can be obtained by looking up a table. The entries of the table contain the residue representations of the numbers, and all the residues of those numbers modulo 2. The size of the table is proportional to the dynamic range M. If M is not big, the size of the redundant modulus 2 table is reasonable. Otherwise, this kind of table is not practical, and the following alternative method should be used.

Given a modulus set $(m_1, m_2, \cdots, m_n)$, whose dynamic range is equal to M, an RNS number X, $(x_1, x_2, \cdots, x_n)$ is corresponding to the modulus set. By the Chinese remainder theorem, X can be converted from its residue number representation by weighted sum such as

$$X = \left| \sum_{j=1}^{n} \hat{m}_j \left| \frac{x_j}{\hat{m}_j} \right|_{m_j} \right|_M \qquad (9.2)$$

with $\hat{m}_j = \frac{M}{m_j}$.

Since $|A|_M$ denotes the least positive residue of A modulo M, Equation (9.2) can be written as

$$X = \sum_{j=1}^{n} \hat{m}_j \left| \frac{x_j}{\hat{m}_j} \right|_{m_j} - rM \qquad (9.3)$$

$$= \frac{M}{m_1} \left| \frac{x_1}{\hat{m}_1} \right|_{m_1} + \frac{M}{m_2} \left| \frac{x_2}{\hat{m}_2} \right|_{m_2} + \cdots + \frac{M}{x_n} \left| \frac{x_n}{\hat{m}_n} \right| - rM \qquad (9.4)$$

with r being an integer.

All the moduli $m_1, m_2, \cdots, m_n$ are odd numbers, therefore $\hat{m}_1, \hat{m}_2, \cdots, \hat{m}_n$ and M are all odd numbers in Equation (9.4). Under this situation $\left|\frac{x_1}{\hat{m}_1}\right|_{m_1}, \left|\frac{x_2}{\hat{m}_2}\right|_{m_2}, \cdots, \left|\frac{x_n}{\hat{m}_n}\right|_{m_n}$ and r will decide the parity of X. Hence, to determine the parity of the number X, all we need to know is the parities of $\left|\frac{x_i}{\hat{m}_i}\right|_{m_i}$ and r. In other words we can extract the least significant bit (LSB) of $\left|\frac{x_i}{\hat{m}_i}\right|_{m_i}$s and r, and "Exclusive OR", $\oplus$, them together, that is,

$$P = LSB\left(\left|\frac{x_1}{\hat{m}_1}\right|_{m_1}\right) \oplus LSB\left(\left|\frac{x_2}{\hat{m}_2}\right|_{m_2}\right) \oplus \cdots \oplus LSB\left(\left|\frac{x_n}{\hat{m}_n}\right|_{m_n}\right) \oplus LSB(r).$$

Here $P = 0$ means that X is an even number, and $P = 1$ means that X is an odd number.

The number $\left|\frac{x_i}{\hat{m}_i}\right|_{m_i}$s can be precalculated, and their LSBs can be stored in a table. Considering that the table storing the parities of $\left|\frac{x_i}{\hat{m}_i}\right|_{m_i}$s is much smaller than the table storing the parities of X, we have reduced the size of the table needed for the parity checking. The next problem is how to find r.

Let $\left|\frac{x_i}{\hat{m}_i}\right|_{m_i} = S_i$, with $i = 1, 2, \cdots, n$, and divide Equation (9.4) by M on both sides. We have

$$\frac{X}{M} = \frac{S_1}{m_1} + \frac{S_2}{m_2} + \cdots + \frac{S_n}{m_n} - r.$$

The integer number r can be found by

$$\frac{S_1}{m_1} + \frac{S_2}{m_2} + \cdots + \frac{S_n}{m_n} - \frac{X}{M}. \qquad (9.5)$$

As we know, a number modulo m_i is less than m_i. Therefore $S_i < m_i$ and $\frac{S_i}{m_i} < 1$. A number X is always less than M, and we can find $\frac{X}{M} < 1$. Obviously r is equal to the integer part of $\sum_i \frac{s_i}{m_i}$, and Equation (9.5) can be rewritten as

$$r = \left\lfloor \frac{S_1}{m_1} + \frac{S_2}{m_2} + \cdots + \frac{S_n}{m_n} \right\rfloor.$$

Note that $\lfloor Y \rfloor$ is the integer part of Y.

Ideally, the binary representation for the fractional part of $\frac{S_i}{m_i}$ has infinite length. However in the physical electronic system it can have only finite length. Suppose that t bits are used to represent the fractional part of $\frac{S_i}{m_i}$. For simplicity we denote $\frac{S_i}{m_i}$ as u_i. Let the rounded value $\hat{u}_i$ be equal to $\lceil 2^t u_i \rceil 2^{-t}$. Since $\hat{u}_i$ is a rounded number of u_i, an error e_i is involved such that

$$\hat{u}_i = u_i + e_i.$$

Taking the summation over i from both sides of the previous equation, we have

$$\sum_i \hat{u}_i = \sum_i u_i + \sum_i e_i. \quad (9.6)$$

Denote $\sum_i \hat{u}_i$, $\sum_i u_i$ and $\sum_i e_i$ as $\hat{U}$, U and e, respectively. Equation (9.6) can be rewritten as

$$U = \hat{U} - e.$$

If we substitute U into Equation (9.5), the equation

$$r = \hat{U} - e - \frac{X}{M}$$

yields. Rearranging the previous equation, we have

$$\hat{U} = r + e + \frac{X}{M}. \quad (9.7)$$

Here we hope that $e + \frac{X}{M} < 1$. In other words, $e < 1 - \frac{X}{M}$.

Since $\frac{X}{M} < 1$, and the smallest difference between 1 and $\frac{X}{M}$ is $\frac{1}{M}$, the integer part of $\hat{U}$ will not be bothered if $e < \frac{1}{M}$, in Equation (9.7). From

$$E = \sum_{i=1}^{n} e_i < \frac{1}{M} \implies ne_i < \frac{1}{M} \implies e_i < \frac{1}{nM},$$

we can choose t (the number of bits in the fractional part of S_i) as

$$t > log_2(nM).$$

In that case the integer r can be represented by the rounded value $\hat{u}_i$ as

$$r = \left\lfloor \sum_{i=1}^{n} \hat{u}_i \right\rfloor.$$

For the calculation of the parity, all we need is the LSB of r. Therefore $t + 1$ bits are needed for $\hat{u}_i$, with 1 bit for the integer part. The value of $\hat{u}_i$ can be precalculated and stored in a table. Since each modulus m_i in RNS is not a large number, the table for storing $\hat{u}_i$ is small. The summation of all $\hat{u}_i$s can be accomplished by using fast multioperand binary adders.

We have already developed several efficient methods for number comparison and overflow detection in order to perform the addition of two positive numbers. In the following chapter, a division algorithm for signed RNS numbers will be presented applying the above derivation.

9.3 DIVISION ALGORITHM

The general division algorithms can be classified into two groups: subtractive algorithms and multiplicative algorithms. The subtractive algorithms recursively subtract the multiple of denominator from the numerator until the difference becomes less than the denominator. The multiple is then the quotient. The multiplicative algorithms compute the reciprocal of the divisor; the quotient is obtained by the multiplication of the reciprocal and the dividend.

Based on the previously described overflow detection and number comparison method, we present in the following first and second subsections a subtractive RNS division algorithm. A multiplicative division algorithm will be presented in the third subsection..

The subtractive algorithm applies sign magnitude arithmetic and binary search. We will first focus on unsigned number division.

9.3.1 Unsigned Number Division

Given two numbers, dividend X and divisor Y, the division in RNS is to find the quotient $Z = \lfloor \frac{X}{Y} \rfloor$.

This algorithm is classified as a subtractive algorithm. Therefore it is necessary to detect the sign in the subtraction and the overflow in the addition.

For simplicities the overflow of the addition of two numbers, X and Y, is denoted as $(X + Y) > M$ in the following equations and algorithms. Given modulus set $(m_1, m_2, \cdots, m_n)$ with dividend $X = (x_1, x_2, \cdots, x_n)$, and divisor $Y = (y_1, y_2, \cdots, y_n)$, we find the quotient Z, where $Z = \lfloor \frac{X}{Y} \rfloor$. The dynamic range M of the RNS is $M = \prod_{i=1}^{n} m_i$.

This algorithm can be divided into three parts. Part I finds 2^k, such that $Y \cdot 2^k \leq X < Y \cdot 2^{k+1}$. Part II finds the difference between 2^k and the quotient. Part III deals with the case $Y \cdot 2^k \leq X < M < Y \cdot 2^{k+1}$ and then go to Part II to find out the difference between 2^k and the quotient.

Part I

>Find the proper 2^k such that $Y \cdot 2^k \leq X < Y \cdot 2^{k+1}$ in the following way. Two variables, lower-bound (LB) and upper-bound (UB), are set to record the range in which the value of the quotient is to be found. LB and UB will dynamically change as the algorithm is executed. In iteration i, $LB = 2^i$ and

$UB = 2^{i+1}$, and we repeatedly compare $(2^i \cdot Y)$ with X and detect whether $(2^{i+1} \cdot Y)$ is greater than M until we find some i, denoted as k, such that $(Y \cdot 2^k) \leq X < (Y \cdot 2^{k+1})$. Then we make the record by setting $\mathcal{LB}_0 = 2^k$ and $\mathcal{UB}_0 = 2^{k+1}$. In each iteration the LB_{i+1} is updated by doubling LB_i, and UB_{i+1} is equal to twice of LB_{i+1}. The following are the equations for finding the upper-bound and the lower-bound.

$$LB_{i+1} = \gamma_{i+1}, \tag{9.8}$$
$$UB_{i+1} = 2 \cdot LB_{i+1} \tag{9.9}$$

with

$$\gamma_{i+1} = \begin{cases} 2 \cdot LB_i, & \text{if } X > Y \cdot LB_i \\ LB_i, & \text{if } X \leq Y \cdot LB_i \end{cases}$$

and $LB_0 = 2^0$.

Suppose that the procedure halts in iteration $(i+1)$ when $X \leq (Y \cdot LB_i)$ is tested. According to Equations (9.8) and (9.9), $LB_{i+1} = LB_i$ and $UB_{i+1} = UB_i$, respectively. Let us define this i to be k and make the following record in $\mathcal{LB}_0$ and $\mathcal{UB}_0$:

$$\begin{cases} \mathcal{LB}_0 = LB_i = 2^k \\ \mathcal{UB}_0 = 2 \cdot LB_i = 2^{k+1}. \end{cases} \tag{9.10}$$

Two cases may occur when the above procedure halts. In one case $(UB_k \cdot Y)$ is smaller than M. Then a binary search starts in Part II. Otherwise go to Part III.

Part II

We have found the upper-bound ($\mathcal{UB}_0$) and the lower-bound ($\mathcal{LB}_0$) from the previous part such that $Y \cdot \mathcal{LB}_0 \leq X < Y \cdot \mathcal{UB}_0$. In this part we perform a binary search to find the difference between $\mathcal{LB}_0$ and the quotient, denoted as QE. k steps are needed to finish this part since 2^k integers exist in the range $[2^k, 2^{k+1})$. Before the binary search we set the initial value QE_0 to be 0. In each step of the binary search we have to compare X with $(Y \cdot \frac{\mathcal{UB}_{j+1} + \mathcal{LB}_{j+1}}{2})$ (for convenience we use another variable "Bounding", B_{j+1}, to denote $\frac{\mathcal{UB}_{j+1} + \mathcal{LB}_{j+1}}{2}$.) If $(X - Y \cdot B_{j+1}) < 0$, set $\mathcal{UB}_{j+1} = B_{j+1}$ and $QE_{j+1} = (2 \cdot QE_j)$. Otherwise set $\mathcal{LB}_{j+1} = B_{j+1}$ and $QE_{j+1} = (2 \cdot QE_j + 1)$. When this procedure is finished we can find the quotient Z to be $Z = \mathcal{LB}_0 + QE_k$. The following are the equations for the binary search.

$$\begin{aligned} B_{j+1} &= \frac{\mathcal{UB}_j + \mathcal{LB}_j}{2} \\ R_{j+1} &= X - Y \cdot B_{j+1} \\ QE_{j+1} &= 2 \cdot QE_j + \delta_{j+1} \\ \mathcal{UB}_j &= \sigma_{j+1} \\ \mathcal{LB}_j &= \theta_{j+1}, \end{aligned}$$

where

$$\delta_{j+1} = \begin{cases} 1, & \text{if } R_{j+1} \geq 0 \\ 0, & \text{otherwise} \end{cases}$$

$$\sigma_{j+1} = \begin{cases} \mathcal{UB}_j, & \text{if } R_{j+1} \geq 0 \\ B_{j+1}, & \text{otherwise} \end{cases}$$

$$\theta_{j+1} = \begin{cases} \mathcal{LB}_j, & \text{if } R_{j+1} < 0 \\ B_{j+1}, & \text{otherwise} \end{cases}$$

and $\delta_0 = 0$, $QE_0 = 0$, $\mathcal{LB}_0 = 2^k$, and $\mathcal{UB}_0 = 2^{k+1}$. The procedure halts when $(j+1) = k$.

Part III

If $(Z \cdot Y) \leq X < M < (Y \cdot 2^{k+1})$, we have to update $\mathcal{UB}_1$ as $\frac{\mathcal{UB}_0 + \mathcal{LB}_0}{2} = \frac{2^k + 2^{k+1}}{2}$, and $\mathcal{LB}_1$ as 2^k. QE_1 is updated as $QE_1 = (2 \cdot QE_0)$. Repeatedly in the $(j+1)$th iteration, update $B_{j+1} = \frac{\mathcal{UB}_j + \mathcal{LB}_j}{2}$ and examine whether $(Y \cdot B_{j+1})$ overflows again.

If there is an overflow, set $\mathcal{UB}_{j+1} = B_{j+1}$ and $QE_{j+1} = (2 \cdot QE_j)$. Continue this procedure until $(Y \cdot B_{j+1})$ does not overflow.

If $(Y \cdot B_{j+1})$ does not overflow and $(X - Y \cdot B_{j+1}) \geq 0$, set $\mathcal{LB}_{j+1} = B_{j+1}$ and $QE_{j+1} = (2 \cdot QE_j + 1)$, and detect overflow again.

If $(Y \cdot B_{j+1})$ does not overflow and $(X - Y \cdot B_{j+1}) < 0$, set $\mathcal{UB}_{j+1} = B_{j+1}$ and $QE_{j+1} = (2 \cdot QE_j)$, and perform the similar operations as defined in the binary search of Part II.

The following equations are applied in the above operations.

$$\begin{aligned}
B_{j+1} &= \frac{\mathcal{UB}_j + \mathcal{LB}_j}{2} \\
R_{j+1} &= Y \cdot B_{j+1} - M \\
R'_{j+1} &= X - Y \cdot B_{j+1} \\
QE_{j+1} &= 2 \cdot QE_j + \delta_{j+1} \\
\mathcal{UB}_{j+1} &= \sigma_{j+1} \\
\mathcal{LB}_{j+1} &= \theta_{j+1},
\end{aligned}$$

where

$$\delta_{j+1} = \begin{cases} 0, & \text{if } (R_{j+1} > 0) \text{ OR } [(R_{j+1} \leq 0) \text{ AND } (R'_{j+1} < 0)] \\ 1, & \text{otherwise} \end{cases}$$

$$\sigma_{j+1} = \begin{cases} B_{j+1}, & \text{if } (R_{j+1} > 0) \text{ OR } [(R_{j+1} \leq 0) \text{ AND } (R'_{j+1} < 0)] \\ \mathcal{UB}_j, & \text{otherwise} \end{cases}$$

$$\theta_{j+1} = \begin{cases} B_{j+1}, & \text{if } [(R_{j+1} \leq 0) \text{ AND } (R'_{j+1} < 0)] \\ \mathcal{LB}_j, & \text{otherwise} \end{cases}$$

$\delta = 0$, $QE_0 = 0$, $\mathcal{LB}_0 = 2^k$, and $\mathcal{UB}_0 = 2^{k+1}$ ($\mathcal{LB}_0$ and $\mathcal{UB}_0$ are from Equation (9.10)).

If $[(Y \cdot \mathcal{B}_{j+1} - M \leq 0) \text{AND} (X - Y \cdot \mathcal{B}_{j+1} < 0)]$, go to Part II and continue the search procedures in Part II. If $(\mathcal{UB}_{j+1} - \mathcal{LB}_{j+1}) = 1$, the search stops. Quotient $Z = \mathcal{LB}_{j+1}$.

The flowchart of Part I to Part III of the algorithm is shown in Figure 9.1.

9.3.2 Signed Number Division

The unsigned RNS division algorithm described in the prior subsection can be further expanded to signed RNS. Sign-magnitude arithmetic and a binary search will be adopted.

With the RNS number comparison technique, the absolute value of the dividend and divisor are used when performing the division calculation, and the overflow in the addition of two numbers is detected. Moreover, the signs of the dividend and the divisor are to be detected, and the negative numbers are to be complemented. After finishing the division on two absolute values, it is necessary to transfer the quotient to the proper representation (positive or negative) in RNS. Given modulus set $\{m_1, m_2, \cdots, m_n\}$ with dividend $X = (x_1, x_2, \cdots, x_n)$ and divisor $Y = (y_1, y_2, \cdots, y_n)$, we find the quotient Z, where $Z = \lfloor \frac{X}{Y} \rfloor$. The dynamic range M of RNS is $M = \prod_{i=1}^{n} m_i$.

This algorithm can be divided into five parts. Part I detects the signs of the dividend and the divisor and converts them to positive numbers. Part II finds 2^k, such that $(Y \cdot 2^k) \leq X \leq (Y \cdot 2^{k+1})$. Part III finds the difference between 2^k and the quotient. Part IV deals with the case $(Y \cdot 2^k) \leq X \leq \frac{M-1}{2} < (Y \cdot 2^{k+1})$ and then go to Part III to find out the difference between 2^k and the quotient. Part V converts the quotient to the proper representation in RNS (positive or negative).

Part I

> The largest number in the positive range of the RNS is $\frac{M-1}{2}$, and for convenience we set a variable $M_p = \frac{M-1}{2}$. Take the absolute value of X and Y and record the signs of them. If the signs of the dividend and divisor are different, then the quotient is negative, and we have to set the sign variable SIGN to 1. SIGN will be used to convert the quotient to a proper form in Part V. Part II to Part IV are very similar to Part I to Part III in the unsigned number division except the initial setting of $M_p = \frac{M-1}{2}$.

Part V

> From Part I the exact quotient may be negative. Therefore if SIGN=1, the absolute value of the found quotient should be complemented.

Suppose the moduli are $m_1 = 3$, $m_2 = 5$ and $m_3 = 7$. Given $X = (2, 1, 1) = -34$ and $Y = (2, 0, 5) = 5$, we are asked to find quotient $Z = \lfloor \frac{X}{Y} \rfloor$.

$M = m_1 \cdot m_2 \cdot m_3 = 105$ and $M_p = \frac{M-1}{2} = 52 = (1, 2, 3)$. The multiplicative inverses of 2, which are used in the calculation of $\frac{UB+LB}{2}$, corresponding to m_1, m_2 and m_3 are $|2^{-1}|_{m_1} = 2$, $|2^{-1}|_{m_2} = 3$, $|2^{-1}|_{m_3} = 4$, respectively. Table 9.1 is referred to for parity checking. The quotient can be calculated within the steps shown in Figure 9.2 with the required variables set in the algorithm.

210   RESIDUE NUMBER OPERATIONS

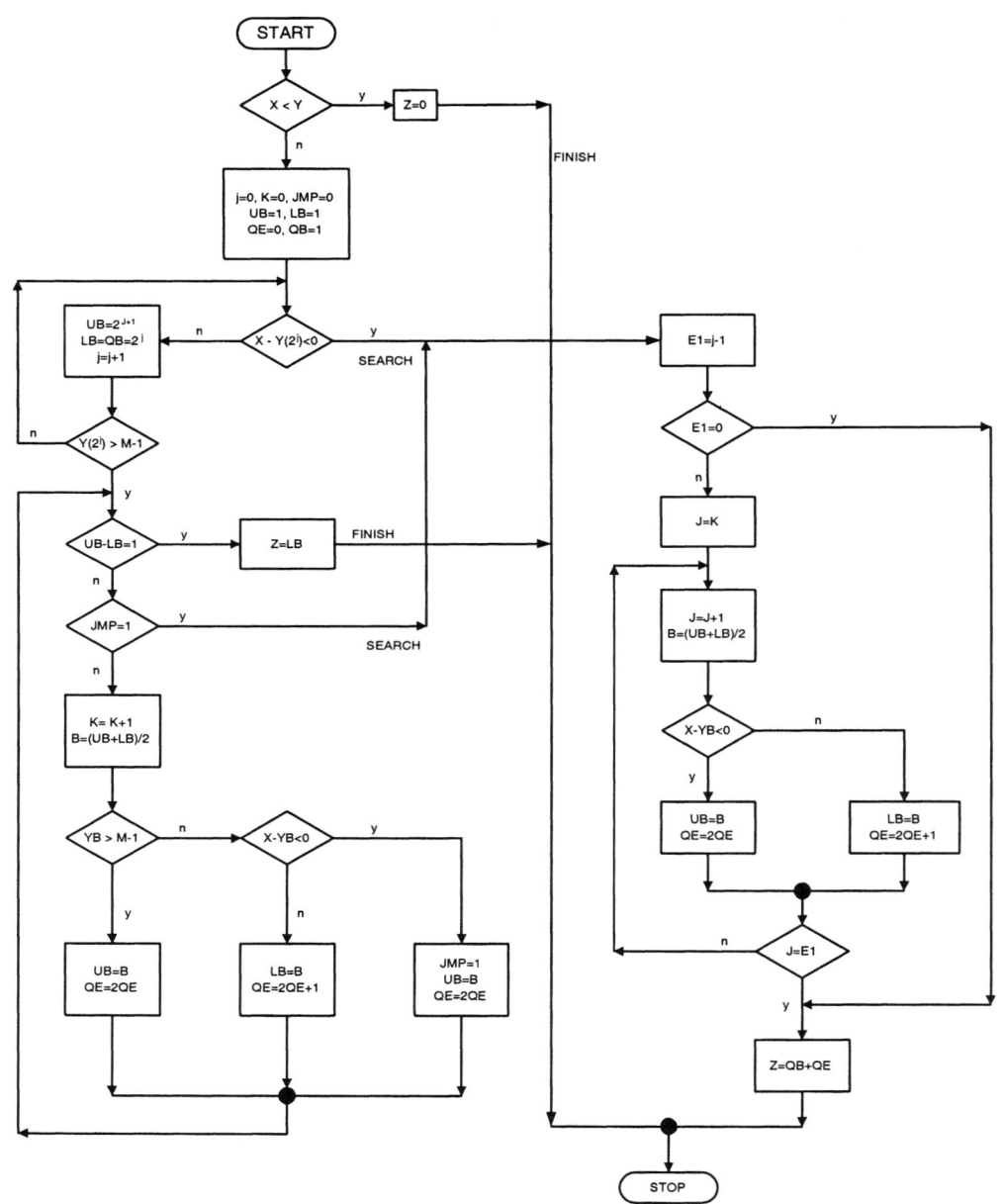

Fig. 9.1: Flowchart of the Unsigned Number Division Algorithm

Step 1.	$S(-34) = 1$, $S(5) = 0$, SIGN=1, COMPLEMENT$(-34) = 34$.	$S[(2,1,1)] = 1$, $S[(2,0,5)] = 0$, SIGN=1, COMPLEMENT$[(2,1,1)] = (1,4,6)$.
Step 2.	$34 > 5 \cdot 2^0$, $j = 0$.	$(1,4,6) > (2,0,5)$, $j = 0$.
Step 3.	$34 > 5 \cdot 2^1$, $j = 1$.	$(1,4,6) > (2,0,5) \cdot (2,2,2) = (1,0,3)$, $j = 1$.
Step 4.	$34 > 5 \cdot 2^2$, $j = 2$.	$(1,4,6) > (1,0,3) \cdot (2,2,2) = (2,0,6)$, $j = 2$.
Step 5.	$5 \cdot 2^3 > 34 > 5 \cdot 2^2$, QB=$2^2$, $J = 0$, UB= 2^3, LB= 2^2, B= $\frac{2^3+2^2}{2} = 6$, QE= $2 \cdot 0 + 1 = 1$, Set LB=B.	$(2,0,5) \cdot (2,2,2) = (1,0,5) > (1,4,6) > (2,0,6)$, QB=$(1,4,4)$, $J = 0$, UB= $(2,2,2) \cdot (2,2,2) \cdot (2,2,2)$, $= (2,3,1)$, LB= $(2,2,2) \cdot (2,2,2) = (1,4,4)$, B= $\frac{(2,3,1)+(1,4,4)}{2} = (0,1,6)$, QE= $(0,0,0) + (1,1,1) = (1,1,1)$, Set LB=B= $(0,1,6)$.
Step 6.	$5 \cdot 2^3 > 34 > 5 \cdot 6$, $J = 1$, UB= 2^3, LB= 6, B= $\frac{2^3+6}{2} = 7$, QE= $2 \cdot 1 + 0 = 2$, Set UB=B.	$(1,0,5) > (1,4,6) > (2,0,5) \cdot (0,1,6) = (0,0,2)$, $J = 1$, UB= $(2,3,1)$, LB= $(0,1,6)$, B= $\frac{(2,3,1)+(0,1,6)}{2} = (1,2,0)$, QE= $(2,2,2) \cdot (1,1,1) + (0,0,0)$, $= (2,2,2)$, Set UB=B= $(1,2,0)$.
Step 7.	$|Z|$ =QB+QE $= 2^2 + 2$	$|Z|$ =QB+QE $= (1,4,4) + (2,2,2) = (0,1,6)$.
Step 8.	SIGN= 1, Z=COMPLEMENT(6), $= -6$.	SIGN= 1, Z=COMPLEMENT$(0,1,6)$, $= (0,4,1)$.

Fig. 9.2: Example of Signed Number Division

9.3.3 Multiplicative Division Algorithm

Multiplicative algorithms use mixed radix number conversion to find the reciprocal of the divisor and to compare numbers. Iteratively the approximate quotient is made closer to the accurate one.

A. Mixed-radix representation

First we describe the mixed radix representation as follows. Given the moduli made of positive pairwise relative prime integers, being ordered as $m_n > \cdots > m_1$. The mixed radix representation of a nonnegative operand X, in the range of $[0, M-1]$, is denoted as

$$X \longleftrightarrow < \hat{x}_n, \hat{x}_{n-1}, \cdots, \hat{x}_1 >,$$

such that

$$X = \hat{x}_n \prod_{i=1}^{n-1} m_i + \hat{x}_{n-1} \prod_{i=1}^{n-2} m_i + \cdots + \hat{x}_2 m_1 + \hat{x}_1, \tag{9.11}$$

with $0 \le \hat{x}_j < m_j$ for all j. For example, given $(m_3, m_2, m_1) = (5, 3, 2)$, the mixed-radix notation $< 4, 0, 1 >$ represents a number

$$X = 4(3 \times 2) + 0(2) + 1 = 25.$$

Table 9.2 lists the mixed-radix digits representing numbers 1 to 13, given $m_3 = 5$, $m_2 = 3$, $m_1 = 2$.

Any positive integer in the range of $[0, M-1]$ has a unique mixed-radix representation. The mixed radix representation of a negative X is defined to be

$$M + X = \hat{x}_n \prod_{i=1}^{n-1} m_i + \hat{x}_{n-1} \prod_{i=1}^{n-2} m_i + \cdots + \hat{x}_2 m_1 + \hat{x}_1, \tag{9.12}$$

where $0 \le \hat{x}_j < m_j$, for all j. In either case, $\hat{x}_j$ is called the jth mixed radix digit of X.

For a set of moduli $m_n, \cdots, m_2, m_1$, the residue representation of a number can be converted to the radix-digit representation of it as follows. From the right-hand side of Equation (9.11) one can see that every term is a multiple of m_1 except the constant $\hat{x}_1$.

So,

$$|X|_{m_1} = \hat{x}_1,$$

indicating that $\hat{x}_1$ is just the first residue digit.

Moving $\hat{x}_1$ from the right-hand side of Equation (9.11) to the left-hand side and dividing both sides by m_1, we have every term on the right-hand side as a multiple of m_2 except the constant $\hat{x}_2$. Hence

$$\left| \frac{X - \hat{x}_1}{m_1} \right|_{m_2} = \hat{x}_2.$$

Table 9.2: Mixed-Radix Digits

Number X	Mixed-radix digits $< \hat{x_3}$	$\hat{x_2}$	$\hat{x_1} >$
0	0	0	0
1	0	0	1
2	0	1	0
3	0	1	1
4	0	2	0
5	0	2	1
6	1	0	0
7	1	0	1
8	1	1	0
9	1	1	1
10	1	2	0
11	1	2	1
12	2	0	0
13	2	0	1

The weights of $\hat{X}_i$ digits are 0 for a_3, 2 for a_2, and 1 for a_1.

Given X, the procedure of converting its residue representation $(x_n, \cdots, x_2, x_1)$ to mixed-radix representation $< \hat{x}_n, \cdots, \hat{x}_2, \hat{x}_1 >$ can be completed by a recursive procedure.

$$X_{i+1} = \frac{X_i - \hat{x}_i}{m_i},$$
$$\hat{x}_{i+1} = X_{i+1}|_{m_{i+1}},$$

where

$$X_1 = X,$$
$$\hat{x}_1 = x_1.$$

Figure 9.3 shows an example for the illustration of the process.

B. Multiplicative division

Let X and Y be the dividend and divisor, respectively, and

$$X \longleftrightarrow < 0, \cdots, 0, \hat{x}_k, \hat{x}_{k-1}, \cdots, \hat{x}_1 >,$$
$$Y \longleftrightarrow < 0, \cdots, 0, \hat{y}_l, \hat{y}_{l-1}, \cdots, \hat{y}_1 >$$

where $\hat{x}_k$ and $\hat{y}_l$ are the most significant nonzero mixed-radix digits of X and Y, respectively. The approximate dividend $\tilde{X}$ is then chosen to be

$$\tilde{X} = \hat{x}_k m_{k-1} \cdots m_2 m_1,$$

214 RESIDUE NUMBER OPERATIONS

$X = 137$

	m_4	m_3	m_2	m_1	
	7	5	3	2	
Residues of X	4	2	2	①	$\hat{x}_1 = 1$
Subtract $\hat{x}_1 = 1$	-1	-1	-1	-1	
$X - \hat{x}_1$	3	1	1	0	
Multiplicative inverse of m_1, $\left|\frac{1}{2}\right|_{m_i}$	4	3	2		multiply
$\left|\frac{X - \hat{x}_1}{2}\right|_{m_i}$	5	3	②		$\hat{x}_2 = 2$
Subtract $\hat{x}_2 = 2$	-2	-2	-2		
$X - \hat{x}_2$	3	1	0		
Multiplicative inverse of m_2, $\left|\frac{1}{3}\right|_{m_i}$	5	2			
$\left|\frac{X - \hat{x}_2}{3}\right|_{m_i}$	1	②			$\hat{x}_3 = 2$
Subtract $\hat{x}_3 = 2$	-2	-2			
$X - \hat{x}_3$	6	0			
Multiplicative inverse of m_2, $\left|\frac{1}{5}\right|_{m_i}$	3				
$\left|\frac{X - \hat{x}_3}{5}\right|_{m_i}$	④				$\hat{x}_3 = 4$

$X = 4(5 \times 3 \times 2) + 2(3 \times 2) + 2(2) + 1$
$ = 120 + 12 + 4 + 1$
$ = 137$

Fig. 9.3: Example of Conversion to Mixed-Radix Representation

which is the dominant part of X, since $\hat{x}_k$ carries the heaviest weight and is the coefficient of the biggest summands among those added toward X in Equation (9.11).

The approximate divisor is chosen to be

$$\tilde{Y} = (\hat{y}_l + 1)m_{l-1} \cdots m_2 m_1.$$

The approximate quotient Z_i can be found by $\tilde{X}/\tilde{Y}$ equal to one of the following values depending on the relationship of k and l.

<u>Case 1</u> $k = l$,
$$\begin{aligned} Z_i &= \frac{\hat{x}_k m_{l-1} \cdots m_2 m_1}{(\hat{y}_l+1)m_{l-1} \cdots m_2 m_1} \\ &= \frac{\hat{x}_k}{\hat{y}_l+1}; \end{aligned}$$

<u>Case 2</u> $k = l + 1$,
$$\begin{aligned} Z_i &= \frac{\hat{x}_k m_l m_{l-1} \cdots m_2 m_1}{(\hat{y}_l+1)m_{l-1} \cdots m_2 m_1} \\ &= \hat{x}_k \frac{m_l}{\hat{y}_l+1}; \end{aligned}$$

<u>Case 3</u> $k > l + 1$,
$$\begin{aligned} Z_i &= \frac{\hat{x}_k \cdots m_{l+1} m_l m_{l-1} \cdots m_2 m_1}{(\hat{y}_l+1)m_{l-1} \cdots m_2 m_1} \\ &= \hat{x}_k \frac{m_{k-1} \cdots m_{l+1} m_l}{(\hat{y}_l+1)}. \end{aligned}$$

Z_i can be found in an iterative procedure. That is,

$$Z_{i+1} = \lfloor \frac{\tilde{X}_i}{\tilde{Y}} \rfloor, \tag{9.13}$$

$$X_{i+1} = X_i - Y Z_{i+1}, \tag{9.14}$$

where $X_0 = X$, the given dividend.

This iterative procedure is continued until either $Z_i = 0$ or $X_i = 0$. It can be proved that Z_i or X_i becomes zero after a finite number of iterations. Suppose this occurs in the rth iteration, from Equation (9.14) we have

$$\begin{aligned} X_1 &= X_0 - Y Z_1 \\ X_2 &= X_1 - Y Z_2 \\ &\cdots \cdots \\ X_{r-1} &= X_{r-2} - Y Z_{r-1} \\ X_r &= X_{r-1} - Y Z_r. \end{aligned}$$

Let's add the above equations up, X_1 on the left of the first equation will cancel that on the right of the second equation, and X_2 on the left of the second equation will cancel that on the right of the third equation, and so forth. Finally we have

$$X_r = X_0 - Y \Big(\sum_{i=1}^{r-1} Z_i + Z_r \Big).$$

On the right-hand side $X_0 = X$. Moving everything else to the left-hand side and dividing both sides by Y, we have

$$\frac{X}{Y} = \sum_{i=1}^{r-1} Z_i + \underbrace{Z_r + \frac{X_r}{Y}}_{Z'_r}$$

$$= \sum_{i=1}^{r-1} Z_i + Z'_r.$$

From

$$Z'_r = Z_r + \frac{X_r}{Y},$$

noticing that x_i is $\tilde{Y}$ dependant, we have

$$Z'_r = \begin{cases} Z_r & \text{if } Z_r \neq 0 \text{ and } X_r = 0, \\ 1, & \text{if } Z_r = 0 \text{ and } X_{r-1} \geq Y \text{ for any } \tilde{Y} \neq Y, \\ 0, & \text{otherwise.} \end{cases}$$

It is obvious that $\frac{X}{Y}$ can be computed by summing up Z_is obtained in various iterations with an adjustment determined at the end. $\lfloor \frac{\hat{x}_k}{\hat{y}_l+1} \rfloor$ and $\lfloor \frac{m_l}{\hat{y}_l+1} \rfloor$ are found by accessing a table which stores the residue representation of all the possible quotients $\lfloor \frac{\alpha}{\hat{y}_l+1} \rfloor$, for $1 \leq \alpha \leq m_n - 1$. This table is indexed by the quantities $\hat{x}_k$ and $\hat{y}_l$ (or by m_l and $\hat{y}_l$.)

The residue codes for all the possible products $m_{k-1} \cdots m_{l+1} m_l$, for $k > l + 1$ are stored in another table indexed by k and l.

REFERENCES

1. A. Avizienis, "Arithmetic Algorithms for Error-Coded Operands," *IEEE Trans. Comp.*, C-22 (June 1973), pp. 567–572.

2. A. Avizienis, "Arithmetic Error Codes: Cost and Effectiveness Studies for Application in Digital System Design," *IEEE Trans. Comp.*, C-20 (Nov. 1971), pp. 1322–1331.

3. D. K. Banerji, "A Novel Implementation for Addition and Subtraction in Residue Number Systems," *IEEE Trans. Comp.*, Vol. C-23, No. 1, Jan. 1974, pp. 106–109.

4. D. K. Banerji and J. A. Brzowjowski, "Sign Dectection in Residue Number Systems," *IEEE Trans. Comp.*, Vol. C-18, No. 4, Apr. 1969, pp. 313–320.

5. D. K. Banerji, T. Y. Cheung and V. Ganesan, "A High-Speed Division Method in Residue Arithmetic," in *5th IEEE Symp. on Comp. Arith.*, 1981, pp. 158–164.

6. W. A. Chren Jr., "A New Residue Number System Division Algorithm," *Computers Math. Applic.*, Vol. 19, No. 7, 1990, pp. 13–29.

7. H. L. Garner, "The Residue Number System," *IRE Trans. Electron. Comp.*, Vol. EC-8, No. 2, June 1959, pp. 140–147.

8. K. Hwang, *Computer Arithmetic: Principles, Architecture and Design*, New York: John Wiley & Sons, 1979, Ch. 9 and 10, pp. 285–357.

9. W. K. Jenkins and B. J. Leon, "The Use of Residue Number Systems in the Design of Finite Impulse Response Digital Filters," *IEEE Trans. Circuits Systems*, Vol. CAS-24, No. 4, 1973, pp. 199–201,

10. G. A. Jullien, "Implementation of Multiplication, Modulo a Prime Number, with Applications to Number Theoretic Transforms," *IEEE Trans. Comp.*, Vol. C-29, No. 10, Oct. 1980, pp. 899–905.

11. G. A. Jullien, "Residue Number Scaling and Other Operations Using ROM Arrays," *IEEE Trans. Comp.*, Vol. C-27, No. 4, Apr. 1978, pp. 325–336.

12. G. A. Jullien and W. C. Miller, "Application of the Residue Number System to Computer Processing of Digital Signals," *Proc. 4th Symp. on Comp. Arith.*, Oct. 1978, IEEE Cat. No. 78CH1412-6C, pp. 220–225.

13. Y. A. Keir, P. W. Cheney and M. Tannenbaum, "Division and Overflow Detection in Residue Number Systems," *IRE Trans. Electron. Comp.*, Vol. EC-11, Aug. 1962, pp. 501–507.

14. E. Kinoshita, H.. Kosako and Y. Kojima, "Floating-Point Arithmetic Algorithms in the Symmetric Residue Number System," *IEEE Trans. Comp.*, Vol. C-23, Jan. 1974, pp. 9–20.

15. E. Kinoshita, H. Kosako and Y. Kojima, "General Division in the Symmetric Residue Number System," *IEEE Trans. Comp.*, Vol. C-22, Feb. 1973, pp. 134–142.

16. M. L. Lin, E. Leiss and B. McInnis, "Division and Sign Detection Algorithm for Residue Number Systems," *Computers Math. Applic.*, Vol. 10, No. 4/5, 1984, pp. 331–342.

17. D. D. Miller, J. N. Polky and J. R. King, "A Survey of Soviet Developments in Residue Number Theory Applied to Digital Filtering," in *26th Midwest Symp. Circuits Systems*, August 1983. (Private Collection, J. S. Chiang.)

18. D. D. Miller and J. N. Polky, "An Implementation of the LMS Algorithm in the Residue Number System," *IEEE Trans. Circuits System*, Vol. CAS-31, May 1984, pp. 452–461.

19. T. R. N. Rao, "Biresidue Error-Correcting Codes for Computer Arithmetic," *IEEE Trans. Comp.*, C-19 (May 1970), pp. 398–402.

20. T. R. N. Rao and A. K. Trehan, "Binary Logic for Residue Arithmetic Using Magnitude Index," *IEEE Trans. Comp.*, Vol. C-19, No. 8, Aug. 1970, pp. 752–757.

21. A. Sasaki, "Addition and Subtraction in the Residue Number System," *IEEE Trans. Electron. Comp.*, Vol. EC-16, No. 9, Apr. 1967, pp. 157–164.

22. A. Sasaki, "The Basis for Implementation of Additive Operations in the Residue Number System," *IEEE Trans. Electron. Comp.*, Vol. EC-17, Nov. 1968, pp. 1066–1073.

23. M. A. Soderstrand, W. K. Jenkins, G. A. Jullien and F. J. Taylor, *Residue Number System Arithmetic Modern Application in Digital Signal Processing*, IEEE Press, N. Y., 1986.

24. N. Szabo and R. Tanaka, *Residue Arithmetic and Its Applications to Computer Technology*, McGraw-Hill, New York, 1967.

25. E. J. Taylor and C. H. Huang, "An Autoscale Residue Multiplier," *IEEE Trans. Comp.*, Vol. C-31, No. 4, Apr. 1982, pp. 321–325.

26. F. J. Taylor, "A VLSI Residue Arithmetic Multiplier," *IEEE Trans. Comp.*, Vol. C-31, June 1982, pp. 540–546.

27. I. M. Vinogradov, *Elements of Number Theory*, Dover, New York, 1954.

28. T. Van Vu, "Efficient Implementations of Chinese Remainder Theorem for Sign Detection and Residue Decoding," *IEEE Trans. Comp.*, Vol. C-34, July 1985, pp. 646–651.

29. S. Waser and M. J. Flynn, *Introduction to Arithmetic for Digital System Designers*, New York, Holt, Rinehart, Winston, 1982, Ch. 5, p. 172.

PROBLEMS

9.1 Given a moduli set (29, 31, 37).
(a) How many numbers can be represented in the residue number system?
(b) Find the residue number representation for $X = 9053$ and $Y = 197$.

9.2 Given the residue number system, and X and Y specified in Problem 9.1, perform $X + Y$ and $X - Y$.

9.3 Given three moduli 3, 5 and 7, represent number $A = 23$ and $B = 4$ in the residue number system. What is the product of $A \times B$?

9.4 For the same moduli set given in Problem 9.3, what number is represented by (0, 0, 2) and (2, 0, 0), respectively?

9.5 For a moduli set (3, 5, 7), compare residue numbers X and Y if the following values are given. The parities of the given numbers can be found in Table 9.1 of the text. Justify the result of your comparison.
(a) $X = (1, 0, 0)$, $Y = (2, 3, 3)$
(b) $X = (1, 0, 1)$, $Y = (2, 3, 4)$
(c) $X = (2, 4, 2)$, $Y = (2, 1, 3)$
(d) $X = 135$, $Y = 233$

9.6 Applying Table 9.1 for and only for the parities of the addends, detect whether any overflow will occur in the following addition $X + Y$. Assume the moduli set is (3, 5, 7) and explain how the detection is performed.
(a) (2, 4, 3) + (2, 1, 5) = (1, 0, 1)
(b) (0, 0, 6) + (2, 0, 1) = (2, 0, 0)
(c) (1, 3, 3) + (2, 4, 4) = (0, 2 0)
(d) (1, 1, 1) + (1, 1, 5) = (2, 2, 6)

9.7 Given a moduli set (2, 3, 5) in the signed RNS system, represent the following subtrahends in their complement form. Perform the subtractions through additions.
(a) $(1, 1, 3) - (1, 1, 2)$
(b) $(1, 0, 4) - (0, 2, 4)$
(c) $(0, 2, 1) - (0, 1, 0)$

9.8 Given a moduli set (2, 3, 5), let the RNS representation of X be (x_1, x_2, x_3) specified as follows. Applying Equations (9.1) and (9.7), find the decimal value of X. Verify that its parity is the same as the calculated by Equation (9.5).
(a) $(x_1, x_2, x_1) = (0, 1, 4)$
(b) $(x_1, x_2, x_1) = (0, 0, 2)$
(c) $(x_1, x_2, x_1) = (1, 0, 3)$

9.9 Given a moduli set (2, 3, 5), represent the unsigned residue number (1,2,1) in a mixed-radix representation.

9.10 For moduli $m_1 = 2$, $m_2 = 3$ and $m_3 = 5$, given $X = (0, 1, 2)$ and $Y = (1, 1, 2)$, perform an unsigned RNS division to find $Z = \lfloor \frac{X}{Y} \rfloor$ following the flow chart given in Fig. 9.1.

10
Operations through Logarithms

10.1 MULTIPLICATION AND ADDITION IN LOGARITHMIC SYSTEMS

Given an X value, from a read only memory (ROM) one can read the value of $log_2 X$ where X is used as the address and $log_2 X$ is the data located at that address. Also, given $log_2 X$ one can find the antilogarithm by reading from ROM the X value. Here $log_2 X$ is used as the address line and X is the data at that address. Note that two separate ROMs are needed for the logarithm and antilogarithm functions, while in the pencil and paper case one table can be shared.

In a logarithmic system multiplication can be completed through addition. First consider unsigned numbers only. Because $log_2(X \times Y) = log_2 X + log_2 Y$, to calculate $X \times Y$, we first find $log_2 X$ and $log_2 Y$. After adding them to get a sum, we find the antilogarithm of the sum. That is knowing $log_2(X \times Y)$, we look for the product of $X \times Y$. Also, in the logarithmic system division can be carried on through subtraction. Because $log_2(X/Y) = log_2 X - log_2 Y$, to calculate X/Y we first find $log_2 X$ and $log_2 Y$. After subtracting $log_2 Y$ from $log_2 X$ we find the antilogarithm of the difference.

Take into consideration the signed numbers. If $P = X \times Y$, P can be represented by $S_P 2^{log_2 |P|}$ where S_P is the sign of product P, and $2^{log_2 |P|} = |P|$. Given $log_2 |P|$, $2^{log_2 |P|}$ denotes the antilogarithm function to find $|P|$ from $log_2 |P|$. Note that sign

$$S_P = S_X \oplus S_Y,$$

and

$$log_2|P| = log_2|X| + log_2|Y|.$$

Similar to the floating point system, $log_2|X|$ and $log_2|Y|$ can be biased. If the biased logarithms are adopted in the above addition, a bias should be deducted after the addition.

Also, if $Q = A/B$, Q can be represented by $S_Q 2^{log_2|Q|}$ where S_Q is the sign of quotient Q, and $2^{log_2|Q|} = |Q|$ denotes the antilogarithm function of finding $|Q|$. Same as in multiplication,

$$S_Q = S_A \oplus S_B,$$

and

$$log_2|Q| = log_2|A| - log_2|B|.$$

If the logarithms are biased, we have to add a bias after the subtraction.

10.2 ADDITION AND SUBTRACTION IN LOGARITHMIC SYSTEMS

Addition and subtraction are difficult in logarithmic systems. There are two approaches to conduct the addition/subtraction. In the first one, addition/subtraction is performed on the true operands in non-logarithmic form. That is, given two inputs represented in logarithm, an antilogarithm operation is needed to find the two true operands. Then add the two true operands or subtract one from the other. Finally find the logarithm of the sum/difference.

The function of finding antilogarithm or logarithm can be performed by table lookup. The given operand can be represented by n bits, then 2^n address lines are needed. The output will be the data located at a particular address, and can be assumed of n bits. Then $2^n \times n$ will be the size for one table and three tables will be needed, two for the antilogarithm function to be used for the two input operands simultaneously and one for the logarithm function for the unit result. The size of the ROM capacity should be $3 \times 2^n \times n$.

In the second approach, the sum/difference of any two operands are pre-calculated and stored in ROM. The addition/subtraction is done by a read from the ROM table.

A straightforward way is to use the two input operands as an address and find from that address the sum/difference. If each input operand is represented by n bits, there will be $2^n \times 2^n$ address lines. Suppose the sum/difference is represented by n bits as well, the size of the table will be $2^{2n} \times n$. For a small $n = 8$ which doesn't lead to a satisfying precision, the capacity of the table is 524,288 bits.

An alternative method to build the table is presented as follows. Let

$$F = A \pm B = \begin{cases} A(1 \pm \frac{B}{A}), & \text{if } |A| > |B| \\ B(\frac{A}{B} \pm 1), & \text{if } |A| \leq |B|. \end{cases} \quad (10.1)$$

ADDITION AND SUBTRACTION IN LOGARITHMIC SYSTEMS

Assume that A and B are both positive for illustration, while the variables can be replaced by their absolute values otherwise. Let's focus on the case of $A > B$ first,

$$\begin{aligned}
log_2 F &= log_2 A \left(1 \pm \frac{B}{A}\right) \\
&= log_2 A + log_2 \left(1 \pm \frac{B}{A}\right) \\
&= log_2 A + log_2 (1 \pm 2^{log_2 \frac{B}{A}}) \\
&= log_2 A + log_2 (1 \pm 2^{log_2 B - log_2 A}) \\
&= log_2 A + log_2 \left(1 \pm 2^{-\overbrace{(log_2 A - log_2 B)}^{\text{easy calculation}}}\right)
\end{aligned}$$

(with underbraces: "antilogarithm table", "easy calculation", "logarithm table")

$(log_2 A - log_2 B)$ can be easily calculated since A and B are inputs in the logarithmic form. Finding $2^{-(log_2 A - log_2 B)}$ is an antilogarithm function to be performed by a table look-up, and the result is ≤ 1. With n bits in the input and p bits in the output fraction, a table of size $2^n \times p$ is sufficient. $1\pm$ the result of antilogarithm is an easy operation, and the output should be sent to a logarithm table.

The size of the logarithm table is an obstacle. It grows exponentially with the number of bits in the operand. Efforts have been made to reduce the size of ROM tables, and the approximation of the logarithm is considered. Let operand X be normalized so that only one nonzero digit is on the left of the radix point, that is, $1 \leq X < 2$. The fraction on the right of the radix point is equal to $x = X - 1$. In Figure 10.1, the straight line depicts the function $Y = X - 1$. Comparing with the curve representing $Y = log_2 X$, one can see that in the range of $[1, 2]$, the straight line is very close to the curve, particularly when X is close to 1 or 2.

Let $A = 28$ and $B = 9$. To find $A \times B$, we first normalize A and B.

$$A = 11100_2 = 1.1100_2 \times 2^4 \qquad (10.2)$$

and

$$B = 01001 = 1.0010 \times 2^3. \qquad (10.3)$$

So,

$$\begin{aligned}
log_2(A \times B) &= log_2 A + log_2 B \\
&= log_2(1.1100_2 \times 2^4) + log_2(1.0010_2 \times 2^3) \\
&= (log_2 1.1100_2 + log_2 2^4) + (log_2 1.0010_2 + log_2 2^3) \\
&\approx 0.1100_2 + 4 + 0.0010_2 + 3 \\
&= 7 + 0.1110_2.
\end{aligned}$$

224 OPERATIONS THROUGH LOGARITHMS

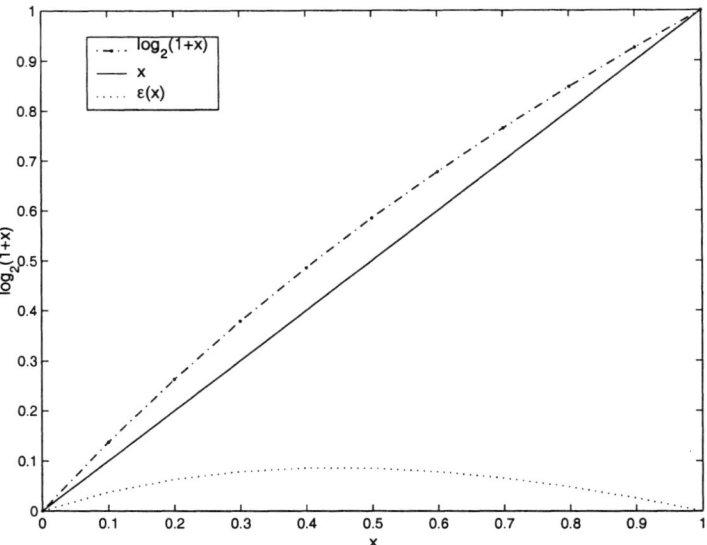

Fig. 10.1: Linear Approximation of $log_2(1+x)$

Because $log_2 X \approx X - 1$, $2^{X-1} \approx X$ for X between 1 and 2,

$$\begin{aligned} A \times B &= 2^{7+0.1110_2} \\ &= 2^7 \times 2^{0.1110_2} \\ &= 128 \times 2^{1.1110_2 - 1} \\ &\approx 128 \times 1.1110_2 \\ &= 240. \end{aligned}$$

The exact value of 28×9 is 252. The error of the above result is $\frac{12}{252} = 4.7\%$. Of course, this is a poor approximation.

The mechanism to perform the above multiplication (or division) is designed as follows (see Figure 10.2). Let the size of registers $\mathcal{A}$ and $\mathcal{B}$ be, for example, 5 bits, and the radix point is assumed to be on the right of MSB, such as x.xxxx with "x" being an arbitrary bit. With its bits contained in such registers, the given number A in the above example is already normalized. No shift is necessary and the exponent in Equation (10.2) is 4. Starting with 0.1001, however, the given number B should be shifted left for one bit position to be normalized, and the exponent in Equation (10.3) is 3. The exponent is represented by three bits of a counter, $x_3 x_2 x_1$ for $\mathcal{A}$ and $y_3 y_2 y_1$ for $\mathcal{B}$ both were initialized as 4 (100_2). The counter will count down for each shift. It remains 4 for the case of A with no shift, and becomes 3 for the case of B, counting down from 4 to 3 after 1 shift. See next section for the formulated details.

The procedure is conducted in the following steps.

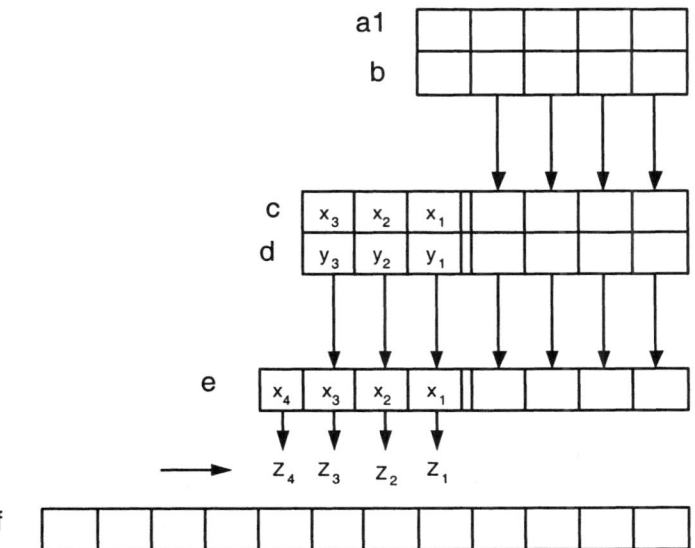

Fig. 10.2: Mechanism for Multiplication (Division) in Binary Logarithms

1. Shift A and B left until their most significant nonzero bits are in the leftmost positions and count down counters $x_3 x_2 x_1$ and $y_3 y_2 y_1$ during shifting.

2. Load bits 4 to 0 of $\mathcal{A}$ into bit positions 4 to 0 of register $\mathcal{C}$, and that of $\mathcal{B}$ into the corresponding bit positions of register $\mathcal{D}$.

3. Perform addition (or subtract) $\mathcal{C} \pm \mathcal{D} \to \mathcal{E}$.

4. Decode $z_4 z_3 z_2 z_1 = Z$ and set bit $Z = 1$ in $\mathcal{F}$. Immediately on the right of that "1", load all the bits from the right of $z_4 z_3 z_2 z_1$ in $\mathcal{E}$. $\mathcal{F}$ now contains the result of $A \times B$ (or $A \div B$).

10.3 REALIZING THE APPROXIMATION

The above normalization procedure can be formulated as follows. Let $b_m b_{m-1} \cdots b_0 . b_{-1} \cdots b_{-p}$ be the binary representation of number B where b_m is the most significant nonzero bit. By factoring out the weight carried by b_m, the rest becomes in between of 1 and 2. That is,

$$B = 2^m + \sum_{i=-p}^{m-1} 2^i b_i$$

$$= 2^m \left(1 + \sum_{i=-p}^{m-1} 2^{i-m} b_i\right)$$
$$= 2^m(1+x),$$

where $i - m < 0$ and $1 \leq 1 + x < 2$. Hence,

$$log_2 B = m + log_2(1+x).$$

In the Taylor series of $log_2(1+x)$, taking only the linear term and let $log_2(1+x) \approx x$, we have

$$log_2 B = m + x.$$

In binary

$$log_2 B = a_n \cdots a_0.b_{m-1} \cdots b_{-p},$$

where $a_n \cdots a_0$ is the binary representation of m, and $b_{m-1} \cdots b_{-p}$ is the binary representation of fraction x, from the lower order part of B in its binary representation. m in the above is referred to as *characteristic* and is actually the number of bits between the most significant nonzero bit and the binary point in number B. One can see later that m can be easily obtained by simple shifting and counting operations.

The error resulted from this method is

$$\varepsilon(x) = log_2(1+x) - x.$$

Clearly $\varepsilon(x)$ is independent of m and depends only on x. Let

$$\frac{\partial \varepsilon(x)}{\partial x} = 0,$$

that is,

$$\frac{log_2 e}{1+x} - 1 = 0,$$

$$x = log_2 e - 1 = 0.44$$

when the maximum error occurs. The maximum error is 0.086. When x=0 or 1, the minimum error occurs. The minimum error is 0. $\varepsilon(x)$ is depicted in Figure 10.1.

Instead of using single-straight-line approximation, several straight lines can be used such as in two, four or eight piecewise-linear approximation where the accuracy can be greatly improved. If four piecewise-linear is selected, for example, the interval [0,1] is divided into four equal subintervals and each of the four straight lines to approximate the curve of logarithm, as depicted in Figure 10.3, is represented by $x + af(x) + b$. The connection of the four segments have been exaggerated in Figure 10.3, with the end point of a segment and that of the subsequent one oppositely

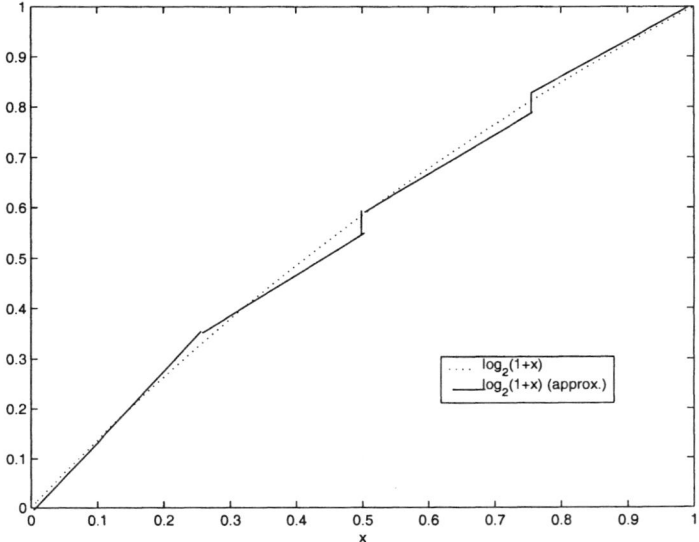

Fig. 10.3: Logarithmic Curve and Four-Straight-Line Approximation

shifted in vertical direction. $f(x)$ can be taken as follows. If for an interval the slope of $log_2(1+x)$ is greater than 1, then $f(x) = x$. If the slope is less than 1, then $f(x) = \bar{x}$ where $\bar{x}$ is the bitwise negating of x. a and b are constants and are selected by trial with a criteria of minimum error, constrained by easy implementation. That is, the coefficients are chosen to be some fraction with the numerator being an integer and denominator being a power of two.

$$log_2(1+x) \approx \begin{cases} x + \frac{5}{16}x, & 0 \leq x < \frac{1}{4} \\ x + \frac{5}{64}x, & \frac{1}{4} \leq x < \frac{1}{2} \\ x + \frac{1}{8}\bar{x} + \frac{3}{128}, & \frac{1}{2} \leq x < \frac{3}{4} \\ x + \frac{1}{4}\bar{x}, & \frac{3}{4} \leq x < 1 \end{cases},$$

where $\bar{x}$ is the 1's complement of x.

The error in this case, $\varepsilon'(x)$, is given in Figure 10.4.

The maximal positive error is 0.008 when $x = 0.44$, and the maximal negative error is -0.006 when $x = 0.25$. The error range is

$$0.008 + 0.006 = 0.014.$$

In the prior single-straight-line case, only positive error occurs and the error range equals 0.086. Hence an improvement of a factor 6 is obtained.

Described below is how the above algorithms can be implemented.

228   OPERATIONS THROUGH LOGARITHMS

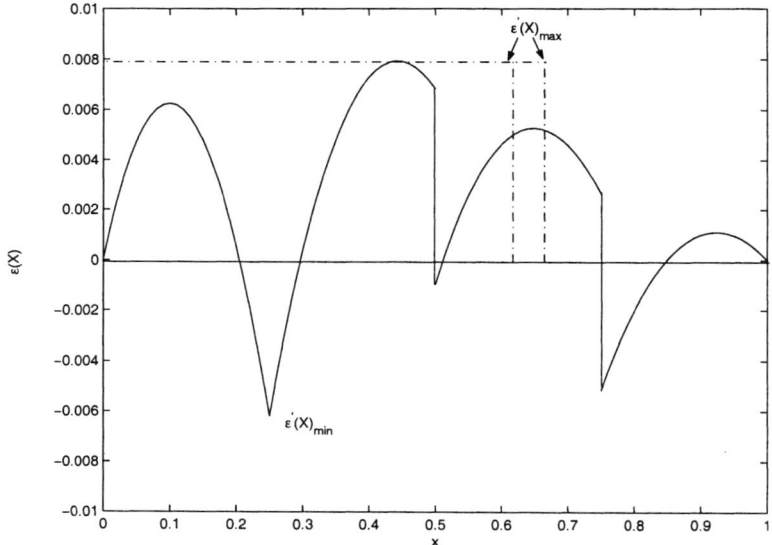

Fig. 10.4: Error of the Four-Straight-Line Approximation

A. Obtaining m

A register R is of k bits long, and a counter C is initialized as k.

The number contained in the register is shifted left until the most significant nonzero bit reaches the end.

$k - m$ shifts will be required, and the counter C is decremented for each shift. At the end $k - (k - m) = m$ is contained in the counter.

B. Corrections

One of the following *corrections* will be added to register R containing x.

$$\begin{cases} \frac{5}{16}x, & \text{if } 0 \leq x < \frac{1}{4} \\ \frac{5}{64}x, & \text{if } \frac{1}{4} \leq x < \frac{1}{2} \\ \frac{1}{8}\bar{x} + \frac{3}{128}, & \text{if } \frac{1}{2} \leq x < \frac{3}{4} \\ \frac{1}{4}\bar{x}, & \text{if } \frac{3}{4} \leq x < 1 \end{cases} \quad (10.4)$$

Seven bits of x are used, and they will be shifted right for p steps to form $\frac{1}{2^p}x$ or $\frac{1}{2^p}\bar{x}$. The coefficients in (10.4) can be composed of $\frac{1}{2^p}$. For example,

$$\frac{5}{16} = \frac{1}{4} + \frac{1}{16},$$
$$\frac{5}{64} = \frac{1}{16} + \frac{1}{64},$$
$$\frac{3}{128} = \frac{1}{64} + \frac{1}{128}.$$

REALIZING THE APPROXIMATION

Table 10.1: Required $\frac{1}{2^P}$s.

X:	0	$\frac{1}{4}$	$\frac{1}{2}$	$\frac{3}{4}$	1
$\frac{1}{4}x$	x				
$\frac{1}{16}x$	x				
$\frac{1}{4}\bar{x}$					x
$\frac{1}{8}\bar{x}$				x	
$\frac{1}{16}$		x			
$\frac{1}{64}$		x	x		
$\frac{1}{128}$				x	

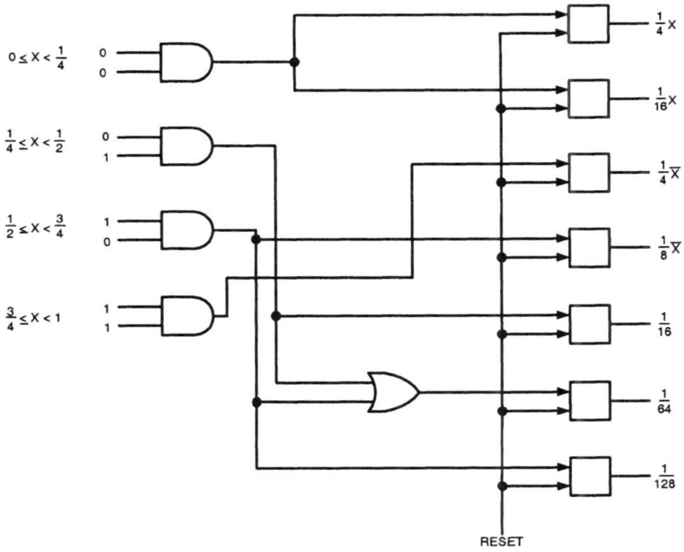

Fig. 10.5: Correction Register

Table 10.1 marked the $\frac{1}{2^P}$s in conjunction with x or $\bar{x}$ needed in different intervals, and Figure 10.5 shows the logic for correct registers. The inputs of each AND gates are from the most significant two bits of x in order to compare x with $\frac{1}{4}$, $\frac{1}{2}$ and $\frac{3}{4}$ and determine which interval x lies.

Figure 10.6 shows how the correction is realized. Stored in R is x. Since the smallest constant is $\frac{1}{128} = \frac{1}{2^7}$, only the most significant 7 bits of x will be effected. These 7 bits are replicated in an auxiliary register. According to the selection made in Figure 10.5, the shifted x or $\bar{x}$ (a fraction of x or $\bar{x}$) will be added to register R.

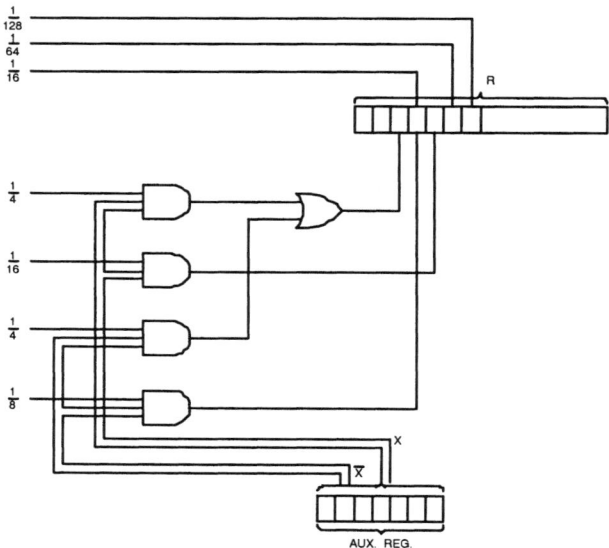

Fig. 10.6: Realization of the Correction

Another method proposes that mean-square error should be minimized. Let $f(x) = x$. The mean-square error over (x_1, x_2), $1 > x_2 \geq x \geq x_1 \geq 0$, is

$$\bar{E}^2 = \frac{1}{x_2 - x_1} \int_{x_1}^{x_2} \{log_2(1+x) - (ax+b)\}^2 dx.$$

To minimize $\bar{E}^2$ with respect to a and b, it is necessary that

$$\frac{\partial \bar{E}^2}{\partial a} = \int_{x_1}^{x_2} -2x\{log_2(1+x) - (ax+b)\}dx = 0$$

$$\frac{\partial \bar{E}^2}{\partial b} = \int_{x_1}^{x_2} -2\{log_2(1+x) - (ax+b)\}dx = 0.$$

Let

$$I_1 = \int_{x_1}^{x_2} log_2(1+x)dx$$

$$= log_2 e\{Y \ln Y - Y\}\Big|_{1+x_1}^{1+x_2}$$

$$I_2 = \int_{x_1}^{x_2} x \, log_2(1+x)dx$$

Table 10.2: Mean-Square Error and Coefficients for Logarithm Approximation

Number of Subintervals	Sub interval	a	b	$\bar{E}^2$	E_{max}
1	1	0.984255	0.065176	0.641074E-3	0.065176
2	1	1.163555	0.021303	0.192903E-4	0.021303
	2	0.827788	0.181567	0.581653E-5	0.010518
4	1	1.285610	0.006243	0.278225E-6	0.006243
	2	1.050957	0.063330	0.387113E-6	0.004141
	3	0.888761	0.143537	0.642476E-6	0.002186
	4	0.770244	0.231857	0.289856E-7	0.002186
8	1	1.359165	0.001681	0.173267E-7	0.001681
	2	1.215426	0.019368	0.550871E-7	0.001371
	3	1.099427	0.048200	0.814192E-7	0.001129
	4	1.003868	0.083914	0.297152E-7	0.000933
	5	0.923414	0.124049	0.130118E-6	0.000794
	6	0.854749	0.166916	0.128720E-6	0.000695
	7	0.806959	0.202033	0.186847E-6	0.001232
	8	0.734065	0.265769	0.150003E-6	0.001186

$$= log_2 e \left\{ \frac{Y^2}{2} ln Y - \frac{Y^2}{4} \right\} \bigg|_{1+x_1}^{1+x_2} - I_1$$

with $1 + x = Y$, then

$$a = \left\{ I_2 - \frac{(x_2 + x_1)I_1}{2} \right\} / \left\{ \frac{x_2^3 - x_1^3}{3} - \frac{(x_2^2 - x_1^2)(x_2 + x_1)}{4} \right\},$$

$$b = \frac{I_1}{x_2 - x_1} - \frac{a(x_1 + x_2)}{2}.$$

For the partition of 1, 2, 4 and 8 subintervals, the coefficients, mean-square error and maximum error are listed in Table 10.2.

To simplify the binary implementation, the linear logarithm equations for a four subdivision realization are given in Table 10.3 with the minimum mean-square coefficients quantitized to seven bits. Smaller number of bits results in only a slightly larger maximum or mean-square error. The maximum error ranges over $-0.00782 < \varepsilon_{max} < 0.00994$. The resulted maximum mean-squared error is $\bar{E}_{max} = 3.33 \times 10^{-6}$.

Given A and hence $log_2 A$, to find $\frac{1}{A}$, a 2's complement operation is needed since $log_2 \frac{1}{A} = -log_2 A$. To find $\sqrt{A}$, a right shift function is needed since $log_2 \sqrt{A} = \frac{1}{2} log_2 A$. To find A^2, a left shift is sufficient since $log_2 A^2 = 2 log_2 A$ and to find the exponent of A, such as A^x, a multiplication is sufficient since $log_2 A^x = x log_2 A$.

Table 10.3: Logarithm Equations

Range	Mantissa
$0 \leq x < 1/4$	$x^* = x + 37x/128 + 1/128$
$1/4 \leq x < 1/2$	$x^* = x + 3x/64 + 1/16$
$1/2 \leq x < 3/4$	$x^* = x + 7\bar{x}/64 + 1/32$
$3/4 \leq x < 1$	$x^* = x + 29\bar{x}/128$

REFERENCES

1. T. A. Brubaker and J. C. Becker, "Multiplication Using Logarithms Implemented with Read-Only Memory," *IEEE Trans. Comp.*, Vol. C-24, No. 8, Aug. 1975, pp. 761–765.

2. M. H. Combet, Van Zonneveld and L. Verbeck, "Computation of the Base Two Logarithm of Binary Numbers," *IEEE Trans. Electron. Comp.*, Vol. EC-14, No. 6, Dec. 1965, pp. 863–867.

3. A. D. Edgar and S. C. Lee, "Focus Microcomputer Number System," *Communications of the ACM*, 22 (Mar. 1979), pp. 166–177.

4. E. L. Hall, D. D. Lynch and S. J. Dwyer III, "Generation of Products and Quotients Using Approximate Binary Logarithms for Digital Filtering Applications," *IEEE Trans. Comp.*, Vol. C-19, No. 2, Feb. 1970, pp. 97–105.

5. N. G. Kingsbury and P. J. W. Rayner, "Digital Filtering Using Logarithmic Arithmetic," *Electron. Lett*, Vol. 7, No. 2, 1971, pp. 56–58.

6. F. S. Lai and C. E. Wu, "A Hybrid Number System Processor with Geometric and Complex Arithmetic Capabilities," *IEEE Trans. Comp.*, 40 (Aug. 1991), pp. 952–962.

7. S. C. Lee and A. D. Edgar, "The Focus Number System," *IEEE Trans. Comp.*, Vol. C-26, No. 11, Nov. 1977, pp. 1167–1170.

8. S. C. Lee and A. D. Edgar, Addendum to "The Focus Number System," *IEEE Trans. Comp.* Vol. C-28, No. 9, Sept. 1979, p. 693.

9. H-Y. Lo and Y. Aoki, "Generation of a Precise Binary Logarithm with Difference Grouping Programmable Logic Array," *IEEE Trans. Comp.*, C-34 (Aug. 1985), pp. 681–691.

10. J. N. Mitchell, Jr., "Computer Multiplication and Division Using Binary Logarithms," *IRE Trans. Electron. Comp.*, Vol. EC-11, No. 4, Aug. 1962, pp. 512–517.

11. E. E. Swartzlander, Jr. and A. G. Alexopoulos, "The Sign/Logarithm Number System," *IEEE Trans. Comp.*, Vol. C-24, No. 12, Dec. 1975, pp. 1238–1242.

12. E. E. Swartzlander, Jr., Comment on "The Focus Number System," *IEEE Trans. Comp.*, Vol. C-28, No. 9, Sept. 1979, p. 693.

13. F. J. Taylor et al., "A 20-Bit Logarithmic Number System Processor," *IEEE Trans. Comp.*, 37 (Feb. 1988), pp. 190–199.

14. L. K. Yu and D. M. Lewis, "A 30-Bit Intergrated Logarithmic Number System Processor," *IEEE J. of Solid-State Circuits*, 26 (Oct. 1991), pp. 1433–1440.

PROBLEMS

10.1 Given the base of logarithm $r = 8$, convert the following numbers into sign-LNS (Logarithmic Number System) representation. Express the LNS part in the 2's complement system with $n = 4$ bits in integer and $k = 2$ bits in fraction.
(a) +64 (b) +0.125 (c) −8 (d) −$\sqrt{8}$

10.2 Given the following sign-LNS representations with base $r = 16$, what number do they represent? Note that one bit is for sign, 4 bits for integer and 4 for fraction in each of the following representations.
(a) $(000000100)_2$ (b) $(011101000)_2$ (c) $(100101000)_2$ (d) $(111111100)_2$

10.3 (a) Based on the representation found in Problem 10.1, compute $X \times Y$.
$X = +64, Y = +0.125$ $X = +0.125, Y = -8$
(b) Based on the representation found in Problems (10.1), compute X/Y.
$X = +64, Y = -8$
$X = -\sqrt{8}, Y = -8$
(c) Based on the representation found in Problem 10.1, compute X^Y.
$X = 8, Y = 0.125$

10.4 Verify the method indicated in Equation (10.1) for the LNS addition/subtraction by performing
(a) $X + Y$ for $X = 512$ and $Y = 128$;
(b) $X - Y$ for $X = 1,024$ and $Y = 256$.
A logarithm ROM table is provided as follows.

X	$\log_2 X$
0.25	−2.0000
0.50	−1.0000
0.75	−0.4150
1.00	0
1.25	+0.3219
1.50	+0.5850
1.75	+0.8074
2.00	+1.0000

10.5 Recalculate the addition and subtraction specified in Problem 10.4, applying the linear approximation of $\log_2(1 + x)$ rather than the ROM table. Give your comments on the error involved.

Hint: Examine every term in the Taylor series for (b).

10.6 Following the procedure similar to that given at the end of Section 10.2, compute $F = A \times B$ with A and B specified as follows.
(a) $A = 35$, $B = 12$. (b) $A = 66$, $B = 10$.

11
Signed-Digit Number Operations

In a signed-digit (SD) number system, carry propagation can be limited to one position to the left during the digit-wise addition and subtraction. The addition time is independent of the word length since the chains of carry-propagations are eliminated.

11.1 CHARACTERISTICS OF SD NUMBERS

In a conventional number representation with an integer radix $r > 1$, each digit is allowed to have exactly r values: $0, 1, \cdots, r - 1$. In an SD representation with the same radix r, each digit is allowed to have more than r values. In the method of addition described below, each digit is allowed to have q different values where

$$r + 2 \leq q \leq 2r - 1. \tag{11.1}$$

Obviously q is more than r. Redundancy in the number representation allows a method of fast addition/subtraction called *totally parallel* addition/subtraction.

In the totally parallel addition/subtraction, the signed-digit representations are required to have a unique representation of zero. Thus the magnitude of allowed digit values may not exceed $r - 1$, since otherwise we could let a digit equal to r and represent zero by $\bar{1}r$ as well as 00.

Some characteristics of the SD numbers are as follows.
Let $A = (a_{n-1} \cdots a_{-k})$ be an SD number, we have

1. $(a_{n-1} \cdots a_{-k}) = 0 \iff a_i = 0 \; \forall i.$

2. $-A = (\bar{a}_{n-1} \cdots \bar{a}_{-k})$, where $\bar{a}_i = -a_i.$

3. Let $p = max\{i | a_i \neq 0\}$, $sign(a_{n-1} \cdots a_{-k}) = sign(a_p)$ for $(a_{n-1} \cdots a_{-k}) \neq 0$.

11.2 TOTALLY PARALLEL ADDITION/SUBTRACTION

Suppose augend $A = (a_{n-1} \cdots a_{-k})$ and addend $B = (b_{n-1} \cdots b_{-k})$. Sum $S = A + B$ is represented by $(s_{n-1} \cdots s_{-k})$.

The addition of two digits is performed in two successive steps. In the first step of addition, the ith bits of augend and addend a_i and b_i are added. A *transfer* digit t_{i+1} and an *interim sum* digit w_i are formed so that

$$a_i + b_i = r \cdot t_{i+1} + w_i. \tag{11.2}$$

t_{i+1} is from the ith digit position to the $(i+1)$th digit position which can carry either a positive or negative value.

In the second step of addition, the sum digit s_i is formed by adding w_i and the t_i from the right adjacent digit position. That is,

$$s_i = w_i + t_i. \tag{11.3}$$

The value of s_i should not exceed the range allowed for the a_i value or b_i value in (11.2). After the first step, there is no carry propagation at any position in the second step.

The block diagram of a totally parallel adder for SD representations is shown in Figure 11.1. Depicted is a 3-bit adder implementing SD number addition/subtraction. Two types of boxes are included. Type (1) computes Equation (11.2) as indicated in computation Step 1. Type (2) computes Equation (11.3) as indicated in computation Step 2 in the preceding section. The digits of A and B are inputs to the type (1) boxes, and the sum digits are obtained from type (2) boxes.

The totally parallel subtraction of b_i from a_i should be performed by a totally parallel addition of the additive inverse of b_i to a_i, that is,

$$a_i - b_i = a_i + \bar{b}_i.$$

Addition of digits a_i and b_i is totally parallel if the following two conditions are satisfied.

1. The sum digit s_i is a function of only a_i, b_i and a *transfer digit* t_i from the $(i-1)$th digit position on right, that is,

$$s_i = f(a_i, b_i, t_i).$$

2. The transfer digit t_{i+1} to the $(i+1)$th position on the left is a function of only a_i and b_i, that is,

$$t_{i+1} = f(a_i, b_i).$$

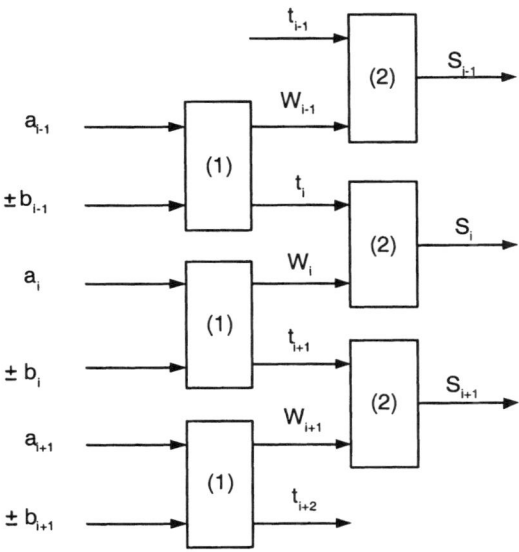

Fig. 11.1: Totally-Parallel Adder in Signed-Digit System

The computation carried on is as follows.

$$
\begin{array}{llll}
0 & a_{n-1} \cdots & a_{-k+1} & a_{-k} \\
0 & b_{n-1} \cdots & b_{-k+1} & b_{-k} \\
0 & w_{n-1} \cdots & w_{-k+1} & s_{-k} \\
\end{array} \Big\} \text{ Step 1}
$$
$$
\begin{array}{llll}
t_n t_{n-1} \cdots & t_{-k+1} & t_{-k} & \\
s_n s_{n-1} \cdots & s_{-k+1} & s_{-k} & \text{Step 2 ,}
\end{array}
$$

where

$$a_i, b_i \in \{-(r-1), \cdots, (r-1)\}. \tag{11.4}$$

s_i should be in the same range as a_i and b_i, that is,

$$|s_i| \leq r - 1. \tag{11.5}$$

11.3 REQUIRED AND ALLOWED VALUES

The required and allowed values of variables are as follows.

$$t_i \in \{-1, 0, 1\},$$

as $|t_i| > 1$ has no practical application for two-digit operations.

$$|w_i| \leq r - 2 \tag{11.6}$$

when t_i is limited as above. What immediately follows is the allowed value of radix r. With the above specified r, w_i is allowed to be $\{-w_{max}, \cdots, -1, 0, 1, \cdots, w_{max}\}$, and

$$t_{i+1} = \begin{cases} 0, & \text{if } |a_i + b_i| \leq w_{max} \\ 1, & \text{if } a_i + b_i > w_{max} \\ -1, & \text{if } a_i + b_i < -w_{max} \end{cases}, \tag{11.7}$$

$$r > 2.$$

For $r = 2$, (11.6) becomes $w_i = 0$ then Equation (11.2) cannot be satisfied if the left-hand side is 1.

Choose $w_i \in \{-(r-2), \cdots, -1, 0, 1, \cdots, r-2\}$ and $r \geq 3$ ($r = 2$ gives the only possible value of w_i as 0, two digit sets are adopted for a totally parallel SD addition/subtraction system.

For an odd r, $r_o \geq 3$, and $\{\frac{-(r_o+1)}{2}, \cdots, -1, 0, 1, \cdots, \frac{r_o+1}{2}\}$ is chosen.
For an even r, $r_e \geq 4$, and $\{-(\frac{r_e}{2}+1), \cdots, -1, 0, 1, \cdots, (\frac{r_e}{2}+1)\}$ is chosen.
From Equation (11.2) we have

$$|a_i|_{max} + |b_i|_{max} = r|t_i|_{max} + |w_i|_{max}.$$

Since $a_i, b_i, \in \{-(r-1), \cdots, (r-1)\}$, we can replace $|a_i|_{max}$ by $r-1$ on the left-hand side, as well as $|b_i|_{max}$. Hence

$$2(r-1) = r|t_i|_{max} + |w_i|_{max}. \tag{11.8}$$

From Equation (11.3) we have

$$|s_i|_{max} = |t_i|_{max} + |w_i|_{max}.$$

Since $s_i \in \{-(r-1), \cdots, r-1\}$, we can replace the left-hand side by $r-1$, and

$$r - 1 = |t_i|_{max} + |w_i|_{max} \tag{11.9}$$

yields.

Solving Equations (11.8) and (11.9), we have

$$\begin{aligned} |t_i|_{max} &= 1, \\ |w_i|_{max} &= r - 2. \end{aligned}$$

There are $2r - 1$ values for a_i, b_i or s_i, from $-(r-1)$ to $(r-1)$. At least r values of w_i are needed to recode all $2r - 1$ values of them, where $w_i = \alpha$ is generated for each pair of values α (when $t_i = 0$) and $-(r - \alpha)$ (when $t_i = -1$), in addition to $w_i = 0$. Furthermore, let the greatest allowed value of w_i be w_{max} and the least allowed value be w_{min}. Anything more than w_{max} is recorded with the help of higher order bit carrying a weight r times more. That is,

$$w_{max} + 1 = r + w_{min},$$

hence

$$w_{max} - w_{min} = r - 1.$$

The value w_{max} is the least when we chose $w_{max} = \frac{r_o - 1}{2}$ for odd radix r_o, and $w_{max} = \frac{r_e}{2}$ for even radix r_e.

Let the set of allowed digit values in SD representation be

$$\{-\alpha, -(\alpha - 1), \cdots, -1, 0, 1, \cdots, (\alpha - 1), \alpha\},$$

where

$$\frac{r_o + 1}{2} \leq \alpha \leq r_o - 1$$

or

$$\frac{r_e}{2} + 1 \leq \alpha \leq r_e - 1$$

with r_o being an odd integer $r_o \geq 3$, and r_e an even integer $r_e \geq 4$. Recall that there should be q different values with q satisfying inequality (11.1), and $\alpha \leq r - 1$ according to inequality (11.5). One can see that for radix 3, the set of allowed digit values is $(-2, -1, 0, 1, 2)$ in which $2r - 1$ values are contained and $\alpha = r - 1$. For radix 4, $(-3, -2, -1, 0, 1, 2, 3)$ is the allowed set satisfying both requirements. For all $r > 4$ there exists more than one set of allowed digit values. For instance for radix 5, $(7 = r + 2) \leq q \leq (2r - 1 = 9)$ and $\alpha \leq (r - 1 = 4)$, hence two sets exist, one with 7 values (-3 to 3) and one with 9 values (-4 to 4). Four sets exist for radix 10, from 13 values (-6 to 6) to 19 values (-9 to 9), as $(12 = r + 2) \leq q \leq (2r - 1 = 19)$ and $\alpha \leq (r - 1 = 9)$.

In the above, when $\alpha = \frac{r_o + 1}{2}$ or $\alpha = \frac{r_e}{2} + 1$ the redundancy is minimal, and when $\alpha = r_o - 1$ or $\alpha = r_e - 1$ the redundancy is maximal. A signed-digit representation may be converted to a canonical form in which the values of all digits a_i are in the chosen set of the values of w_i; that is, no more transfer digits can be formed. To put an m-digit number into the canonical form, a maximum of m additions of the number to zero may be required.

11.4 MULTIPLICATION AND DIVISION

Multiplication and division are executed as sequences of additions/subtractions and shifts.

Given multiplicand A and $m+1$-bit multiplier $B = b_0.b_1 \cdots b_m$ in radix $r \geq 3$ signed-digit representations, the addition of two signed-digit numbers is performed as the totally parallel addition of all corresponding digits a_i and b_i according to Equations (11.2) and (11.3). The product P_{m+1} is formed in the following recursive process:

$$P_{j+1} = \frac{1}{r}(P_j + A \cdot b_{m-j}) \text{ for } j = 0, 1, 2, \cdots, m-1. \quad (11.10)$$

Here P_{j+1} is a partial product, P_0 is assumed 0, and b_{m-j} is the multiplier digit sensed during the jth step of multiplication.

Note that if $|P_j| \leq |A|_{max}$ and $|A| \leq |A|_{max}$, we have

$$|P_j + A \cdot b_{m-j}| \leq |A|_{max}(1 + b_{m-j}), \quad (11.11)$$

where $|A|_{max}$ is the maximum allowed magnitude of a number. Since $|b_{m-j}| \leq r-1$ always holds, after plus 1 and multiply by $\frac{1}{r}$, we have

$$\frac{1}{r}|A|_{max}(1 + b_{m-j}) \leq |A|_{max}.$$

From (11.11),

$$\frac{1}{r}|P_j + A \cdot b_{m-j}| \leq |A|_{max};$$

that is,

$$P_{j+1} \leq |A|_{max}$$

due to (11.10). Table 11.1 shows a numerical (Radix 10) example. Multiplicand $A = 0.\bar{3}26$, and multiplier $B = 1.\bar{2}1\bar{5}$. The value of A is -0.274 and that of B is 0.805. The product $P_4 = AB = 0.\bar{3}8\bar{1}43$ with a value of it being -0.22057.

Signed-digit division is performed as a sequence of additions/subtractions and left shifts. The representation of quotient digits in this method may be redundant, while the redundancy allows a flexible selection of quotient digits. In a signed-digit representation the sign of a partial remainder is not readily available which is required by Robertson's restoring or nonrestoring division. The magnitude of the partial remainder, however, may be estimated from the inspection of a few most significant digits. Given dividend N and divisor D, the quotient digits q_js are generated by the following recursive process:

$$\begin{aligned} R_{j+1} &= r \times R_j - q_{j+1} \times D, \text{ for } j = 0, 1, 2, \cdots, m-1, \\ R_0 &= N - q_0 D. \end{aligned}$$

Here R_j is a partial remainder, R_m is the (final) reminder, and q is the quotient of $(m+1)$ digits.

During each step of division, the quotient digit q_{j+1} must be chosen which has a value such that the next partial remainder R_{j+1} is within the same allowed range as

Table 11.1: Example for SD Multiplication

Step	Digit b_{m-j}	Variable	Quantity	Operation
$j=0$	$b_3 = \bar{5}$	P_0 Ab_3	0.000 1.4$\bar{3}$0	add
		$P_0 + Ab_3$ P_1	1.4$\bar{3}$0 0.14$\bar{3}$	shift right
$j=1$	$b_2 = 1$	Ab_2	0.$\bar{3}$26	add
		$P_1 + Ab_2$ P_2	0.$\bar{2}$63 0.0$\bar{2}$63	shift right
$j=2$	$b_1 = \bar{2}$	Ab_1	0.6$\bar{5}\bar{2}$	add
		$P_2 + Ab_1$ P_2	0.6$\bar{7}$43 0.06$\bar{7}$43	shift right
$j=2$	$b_0 = 1$	Ab_0	0.$\bar{3}$26	add
		P_1	0.$\bar{3}$8$\bar{1}$43	end

Table 11.2: Example for SD Division

Step	Variable	Quantity	Test	Operation	q_{j+1}
$j = -1$	N	$0.2\bar{1}6\bar{3}45$	$0.204 < T$	shift	$q_0 = 0$
$j = 0$	rR_0 D	$\bar{2}.\bar{1}6\bar{3}450$ $1.\bar{3}07$	$2.043 > T$	add D	
	$rR_0 + D$ D	$\bar{1}.\bar{4}64450$ $1.\bar{3}07$	$1.336 > T$	add D	
	$rR_0 + 2D$ D	$0.\bar{7}71450$ $1.\bar{3}07$	$0.629 > T$	add D	
	R_1	0.078450	$0.078 < T$	shift	$q_1 = \bar{3}$
$j = 1$	rR_1 $-D$	0.78450 $\bar{1}.30\bar{7}$	$0.784 > T$	add $-D$	
	R_2	$0.08\bar{3}50$	$0.077 < T$	shift	$q_2 = 1$
$j = 2$	rR_2 $-D$	$0.8\bar{3}50$ $\bar{1}.30\bar{7}$	$0.775 > T$	add $-D$	
	R_3	$0.1\bar{3}\bar{2}$	$0.068 < T$	end	$q_3 = 1$

R_j. This range is a function of the magnitude of divisor D. If dividend N is not in this range, the choice of $|q_0| = 1$ must bring R_0 into the allowed range, otherwise the quotient overflows. Every quotient digit q_{j+1} is assigned the value which satisfies the condition

$$|rR_j - Dq_{j+1}| \leq c|D|, \tag{11.12}$$

where c is the range test constant whose allowed range is determined by the choice of the allowed values of the quotient digit. $c = \frac{1}{2}$ is the most practical choice within the allowed range. If the representation of the quotient digits is redundant (q_{j+1} is assumed more than r values), the comparison in (11.12) may be inexact; that is, it is sufficient to perform the comparison between truncated values of $|rR_j - Dq_{j+1}|$ and $|D|$. To facilitate the comparison, the divisor D must be standardized before division, otherwise very great precision of comparison is required when $|D|$ is small.

There are two considerations for the choice of the values allowed for quotient digits q_j. First, if q_j is assumed a least possible number, that number of additions will be required for one division. Second, minimal redundancy in representation of q_j is necessary to allow an inexact selection of quotient digits. In minimal redundancy, the values of q_j range from $-\frac{1}{2}r_e$ to $\frac{1}{2}r_e$ (a total of $r+1$ values) for even radices $r_e \geq 4$, and from $-\frac{1}{2}(r_o + 1)$ to $\frac{1}{2}(r_o + 1)$ (a total of $r+2$ values) for odd radices $r_o \geq 3$. To determine q_j out of these values, it is sufficient to compare the first four digits.

In particular, D is repetitively added to/subtracted from the shifted partial remainder rR_j, until the following condition is detected indicating R_{j+1} has been generated:

$$\left|\sum_{i=0}^{3} n_i r^{-i}\right| \leq \left|\frac{1}{2}\sum_{i=0}^{3} d_i r^{-i}\right|, \tag{11.13}$$

where $n_i (i = 0, 1, 2, 3)$ are the first four digits in the accumulator register containing the dividend N when the division began, and the partial remainders later. The value of q_{j+1} is equal to the number of additions or subtractions required to generate R_{j+1}. If (11.13) is satisfied right after the left shift of R_j, $q_{j+1} = 0$ is the found value. A numerical (Radix 10) example is shown in Table 11.2. Dividend $N = 0.\bar{2}1\bar{6}34\bar{5}$, and divisor $D = 1.\bar{3}0\bar{7}$. The value of N is -0.204255 and that of D is 0.707. The test quantity is

$$T = \left|\frac{1}{2}\sum_{i=0}^{3} d_i r^{-i}\right|$$

with the value $T = 0.3535$. The results are $Q = 0.\bar{3}11$ and $R_3 = 0.1\bar{3}\bar{2}$, representing a quotient value equal to -0.289 and a final remainder 0.000068, respectively.

REFERENCES

1. A. Avizienis, "Binary-Compatible Signed-Digit Arithmetic," *Proc. AFIPS Fall Joint Comp. Conf.*, 1964, pp. 663–671.

2. A. Avizienis, "A Study of Redundant Number Representations for Parallel Digital Computers," Ph.D. Thesis, University of Illinois, Urbana, May 1960.

3. A. Avizienis, "Signed-Digit Number Representations for Fast Parallel Arithmetic," *IRE Trans. Electron. Comp.*, Vol. EC-10, No. 3, Sept. 1961, pp. 389–400.

4. B. Parhami, "Generalized Signed-Digit Number Systems: A Unifying Framework for Redundant Number Representations," *IEEE Trans. Comp.*, 39 (Jan. 1990) pp. 89–98.

5. C. Tung, "Signed-Digit Division Using Combinational Arithmetic Nets," *IEEE Trans. Comp.*, Vol. C-19, No. 8, Aug. 1970, pp. 746–748.

PROBLEMS

11.1 (a) Given a digit set $\{\bar{1}, 0, 1\}$, find all the possible signed-digit (SD) representations of $(-5)_{10}$ with $n = 4$ and $k = 0$.
(b) Among those representations you obtained in (a), which one is the minimal SD representation?

11.2 Given $r = 8$ and $\alpha = 5$, convert the following numbers into equivalent SD numbers.
(a) $(725)_8$ and $(-725)_8$. (b) $(671)_8$ and $(-671)_8$.

11.3 Applying the method indicated in Figure 11.1, perform addition $A + B$ with the following radix 10 numbers. First, find w_is and t_is in parallel, then find all the s_is simultaneously assuming $w_{max} = 5$.
(a) $A = 03\bar{1}\bar{1}, B = \bar{1}43\bar{5}$. (b) $A = 1\bar{2}\bar{2}1, B = 11\bar{4}3$.

11.4 Perform subtraction $A - B$ with the same numbers given in Problem 11.3. Similar requirements and restrictions apply.

11.5 How many sets of redundant signed digits can be used to represent a radix $r = 8$ number. What are they?

11.6 Represent each of the following numbers in radix 8 SD form with the different sets of redundant signed digits you found in Problem 11.5.
(a) 267_8. (b) 501_{10}.

11.7 Perform the SD multiplication of $A \times B$ where $A = 0.\bar{4}62$ and $B = 1.\bar{2}1\bar{5}$ are radix 10 numbers. For each step, list the operations performed and the partial product generated.

11.8 Perform the SD division of N/D where $A = 0.\bar{2}36\bar{4}51$ and $B = 1.\bar{3}1\bar{5}$ are radix 10 numbers. For each step, list the operations, the test condition, and the partial remainder and quotient obtained.

Index

Add-and-shift approach, 78
   accumulator(AC), 78
   auxiliary register(AX), 78
   multiplier register(MR), 78
Addition/subtraction, 40
   addend, 40
   augend, 40
   minuend, 40
   ripple carry adder, 44
   subtrahend, 40
   summand, 40
Arithmetic logic unit, 29
Arithmetic shift, 87
Array divider, 167
   carry-lookahead cellular array divider, 173
   non-restoring cellular array divider, 171
   restoring cellular array divider, 167
Baugh-Wooley's array, 114
Bi-section array multiplier, 119
Bit partitioned adder, 69
   column adder, 71
Block carry lookahead adder (BCLA), 62
   block carry generate, 62
   block carry propagate, 62
Carry-completion sensing adder (CCSA), 56
   carry propagation length, 58
Carry-lookahead adder (CLA), 61
   carry generation, 61
   carry propagation, 61
Carry-save adder (CSA), 66

carry-save register, 66
Conditional-sum adder, 54
Conventional radix number, 2
Convergence division, 152
Dependent carry (DC), 56
Diminished radix complement, 8
Division by divisor reciprocation, 157
Fixed-radix number system, 2
   binary, 4
   hexadecimal, 4
   octal, 4
Floating-point number representation, 15
   bias, 16
   double precision representation, 20
   exponent, 15
   mantissa, 15
   normalization, 15
   order-of-magnitude zero, 184
   significand, 15
   underflow, 19
Full-adder(FA), 32
Half-adder(HA), 32
Independent carry (IC), 56
Indirect multiplication, 81
   post-complement, 81
   pre-complement, 81
Least significant digit (LSD), 3
Logarithmic number system (LNS), 23
Mixed-radix number, 2
Modular structure, 120

additive multiply modules, 123
non-additive multiply modules, 123
Most significant digit (MSD), 3
Non-overlapped multiple-bit scanning, 90
Normalized fraction, 7
One's complement addition
   end-around carry, 46
Overflow, 43
Overlapped multiple bit scanning
   Booth's algorithm, 93
   canonical multiplier recoding, 95
   canonical signed-digit vector, 95
   string property, 90
Overlapped multiple bit scanning, 90
Pezaris array multiplier, 117
Radix, 9
Residue number system (RNS), 22
   moduli set, 22
   multiplicative division algorithm, 212
   multiplicative inverse, 203
   subtractive RNS division algorithm, 206
Robertson's signed number multiplication, 87
Rounding, 193
   guard bit, 193
   round bit, 193
   round-to-nearest, 190
   round-up, 190
   sticky bit, 193
   truncate, 190
Sign-magnitude addition, 48
   post-complement, 49
   pre-complement, 48
Sign-magnitude, 8
Signed-digit number system, 12
   minimal signed-digit representation, 14
   totally parallel addition/subtraction, 235
Subtract-and-shift approach, 136
   binary non-restoring division, 141
   binary restoring division, 138
   high-radix restoring division, 144
   modified SRT division, 147
   Robertson's high-radix division, 147
   SRT division, 146
Tri-section array multiplier, 119
Wallace tree, 103